Key National Education Indicators

Workshop Summary

Alexandra Beatty and Judith A. Koenig, *Rapporteurs*

Steering Committee on Workshop on Key National Education Indicators

Board on Testing and Assessment
and
Committee on National Statistics

Division of Behavioral and Social Sciences and Education

NATIONAL RESEARCH COUNCIL
OF THE NATIONAL ACADEMIES

THE NATIONAL ACADEMIES PRESS
Washington, D.C.
www.nap.edu

KH

THE NATIONAL ACADEMIES PRESS   500 Fifth Street, NW   Washington, DC 20001

This study was supported by Contract No. 11-98169-000-USP between the National Academy of Sciences and the John D. and Catherine T. MacArthur Foundation. The study was also supported by the Carnegie Corporation of New York, the Bill & Melinda Gates Foundation, and by the Presidents' Fund of the National Research Council. Any opinions, findings, conclusions, or recommendations expressed in this publication are those of the author(s) and do not necessarily reflect the views of the organizations or agencies that provided support for the project.

International Standard Book Number-13: 978-0-309-26121-0
International Standard Book Number-10: 0-309-26121-X

Additional copies of this report are available from the National Academies Press, 500 Fifth Street, NW, Keck 360, Washington, DC 20001; (800) 624-6242 or (202) 334-3313; http://www.nap.edu.

Suggested citation: National Research Council. (2012). *Key National Education Indicators: Workshop Summary.* Steering Committee on Workshop on Key National Education Indicators, A. Beatty and J.A. Koenig, Rapporteurs. Board on Testing and Assessment and Committee on National Statistics, Division of Behavioral and Social Sciences and Education. Washington, DC: The National Academies Press.

3/15/13

THE NATIONAL ACADEMIES
Advisers to the Nation on Science, Engineering, and Medicine

The **National Academy of Sciences** is a private, nonprofit, self-perpetuating society of distinguished scholars engaged in scientific and engineering research, dedicated to the furtherance of science and technology and to their use for the general welfare. Upon the authority of the charter granted to it by the Congress in 1863, the Academy has a mandate that requires it to advise the federal government on scientific and technical matters. Dr. Ralph J. Cicerone is president of the National Academy of Sciences.

The **National Academy of Engineering** was established in 1964, under the charter of the National Academy of Sciences, as a parallel organization of outstanding engineers. It is autonomous in its administration and in the selection of its members, sharing with the National Academy of Sciences the responsibility for advising the federal government. The National Academy of Engineering also sponsors engineering programs aimed at meeting national needs, encourages education and research, and recognizes the superior achievements of engineers. Dr. Charles M. Vest is president of the National Academy of Engineering.

The **Institute of Medicine** was established in 1970 by the National Academy of Sciences to secure the services of eminent members of appropriate professions in the examination of policy matters pertaining to the health of the public. The Institute acts under the responsibility given to the National Academy of Sciences by its congressional charter to be an adviser to the federal government and, upon its own initiative, to identify issues of medical care, research, and education. Dr. Harvey V. Fineberg is president of the Institute of Medicine.

The **National Research Council** was organized by the National Academy of Sciences in 1916 to associate the broad community of science and technology with the Academy's purposes of furthering knowledge and advising the federal government. Functioning in accordance with general policies determined by the Academy, the Council has become the principal operating agency of both the National Academy of Sciences and the National Academy of Engineering in providing services to the government, the public, and the scientific and engineering communities. The Council is administered jointly by both Academies and the Institute of Medicine. Dr. Ralph J. Cicerone and Dr. Charles M. Vest are chair and vice chair, respectively, of the National Research Council.

www.national-academies.org

iii

Preface

The education system in the United States is continually challenged to adapt and improve, in part because its mission has become far more ambitious than it once was. At the turn of the 20th century, fewer than one-tenth of students enrolled were expected to graduate from high school, and it was only in the 1960s that the expectation that all students would graduate became widespread (National Research Council, 2001). Today, most people expect schools to prepare all students to succeed in postsecondary education and to prosper in a complex, fast-changing global economy. Goals have broadened to include not only rigorous benchmarks in core academic subjects, but also technological literacy and the subtler capacities known as 21st century skills.

As these changes have taken place, education research has become increasingly clear in pointing to some of the key elements that make teaching and learning successful, and educators and leaders are under intense pressure to apply this knowledge every day in improving schools.

These high expectations mean that the American public has pressing questions about how well students are learning and how well schools are doing. Existing measures reveal some uncomfortable though important truths about gaps in student achievement and schools that are not succeeding, and they also highlight areas of considerable strength. But existing measures do not provide answers to all the questions that have been raised.

To identify the most important measures for education and other issues and provide quality data on them to the American people, Congress has authorized the creation of a Key National Indicators System (KNIS). This system would be a single web-based information source designed to help policy makers and the public better assess the position and progress of the nation across a wide range of areas.

Identifying the right set of indicators for each area is not a small challenge. To serve their purpose of providing objective information that can encourage improvement and innovation, the indicators need to be valid and reliable but they also need to capture our aspirations for education. They need to take advantage of wisdom that has been developed over many years of research and practice. They also must anticipate the changes taking place as innovation and technology reshape education in this country and around the world. They need to focus on what is most important but also to capture sophisticated aspects of intellectual development, and the complex interactions among individuals, families, communities, school systems, and public policy. Perhaps most important, they need to effectively communicate specific information about how well the U.S. education system is doing.

This report describes a workshop, planned under the aegis of the Board on Testing and Assessment and the Committee on National Statistics, and funded by the John D. and Catherine T. MacArthur Foundation, the Carnegie Corporation of New York, the Bill & Melinda Gates Foundation, and the Presidents' Fund of the National Research Council.

The workshop was designed as an opportunity for a group with extensive experience in research, public policy, and practice could begin the process of sorting through possible indicators and considering possibilities for developing new ones. The goal of the workshop was not to make a final selection of indicators, but to make an important first step by clearly identifying the parameters of the challenge. On behalf of the steering committee, I hope this report will be the stimulus for further thinking on this important issue.

Many people contributed to the success of this workshop. We first thank the sponsors, the John D. and Catherine T. MacArthur Foundation, the Carnegie Corporation of New York, and the Bill and Melinda Gates Foundation, for their support of this work. We also thank the Presidents' Committee of the National Research Council. for its support. We sincerely appreciate all of the insights provided by Chris Hoenig, senior advisor to the presidents of the National Academies and CEO of the State of the USA.

The steering committee also thanks the scholars who wrote papers and made presentations at the workshop: Judith Alamprese, Abt Associates; Steven Barnett, Rutgers University; Margaret Burchinal, University of North Carolina at Chapel Hill; Kevin Dougherty, Columbia University; Emerson Elliott, National Council for Accreditation of Teacher Education; Ronald Ferguson, Harvard University; Eugene García, Arizona State University; Brian Gill, Mathematica; Patricia Graham, Harvard University; Joseph Kahne, Mills College; Laura Perna, University of Pennsylvania; Robert Pianta, University of Virginia; Lashawn Richburg-Hayes, MDRC; Donald Roberts, Stanford University; Sue Sheridan, Nebraska University; Marshall S. Smith, Carnegie Foundation for the Advancement of Teaching; Elizabeth Stage, Lawrence Hall; David Stern, University of California at Berkeley; Patrick Terenzini, Pennsylvania State University; Bill Tierney, University of Southern California; and Deborah Vandell, University of California at Irvine. We are especially indebted to Emerson Elliott and Norman Bradburn, with the National Opinion Research Center, for sharing their papers and providing helpful input during all stages of this project. We also thank John Ralph, with the U.S. Department of Education, for providing historical information about prior efforts to identify education indicators.

The steering committee also thanks the National Research Council staff who worked directly on this project. Judith Koenig, study director, served tirelessly in helping to assemble panelists, structure the order of presentations, and define the issues for discussion. Without her persistent efforts, the workshop could not have happened. Stuart Elliott, BOTA director, provided insightful contributions in helping to formulate the design of the workshop and make it a reality. Kelly Iverson, senior project assistant, provided her deft organizational skills and careful attention to detail to help to ensure the success of the workshop. We particularly wish to recognize Alix Beatty, senior program officer, for her superb writing skills and ability to translate workshop presentations and discussions into a coherent, readable report.

Finally, as chair of the steering committee, I thank the members for their dedication and outstanding contributions to this project. They gave generously of their time in planning the workshop and participated actively in workshop presentations and discussions. Their varied experiences and perspectives contributed immeasurably to the success of the project and made them a delightful set of colleagues for this work.

This workshop summary has been reviewed in draft form by individuals chosen for their diverse perspectives and technical expertise, in accordance with procedures approved by the Report Review Committee of the National Research Council. The purpose of this independent review is to provide candid and critical comments that will assist the institution in making its published report as sound as possible and to ensure that the report meets institutional standards for objectivity, evidence, and responsiveness to the charge. The review comments and draft manuscript remain confidential to protect the integrity of the process. We thank the following individuals for their review of this report: W. Steven Barnett, National Institute for Early Education Research, Rutgers University; Rolf Blank, Research, Education Indicators, Council of Chief State School Officers; Christopher T. Cross, Chairman's Office, Cross & Joftus, LLC; Kevin J. Dougherty, Teacher's College, Columbia University; Ronald F. Ferguson, Kennedy School of Government, Harvard University; James P. Gee, Division of Education Leadership and Innovation, Arizona State University; Brian Gill, Senior Fellow, Mathematica Policy Research; and Marshall S. Smith, Education Policy, The Carnegie Foundation for the Advancement of Teaching.

Although the reviewers listed above provided many constructive comments and suggestions, they were not asked to endorse the content of the report nor did they see the final draft of the report before its release. The review of this report was overseen by Lorraine McDonnell, Department of Political Science, University of California, Santa Barbara. Appointed by the National Research Council, she was responsible for making certain that an independent examination of this report was carried out in accordance with institutional procedures and that all review comments were carefully considered. Responsibility for the final content of this report rests entirely with the authors and the institution.

David Breneman, *Chair*
Steering Committee on Workshop on Key National Education Indicators

Contents

1
Introduction

Each month the American public receives a report on the economic health of the nation. The report provides primary information, such as the unemployment rate, the number of new jobs created, and the number of new applications for unemployment are up or down from the previous month. The key indicators that are highlighted each month are only a few from among a vast cornucopia of economic data that might be reported. Together they provide a quick and reasonable overview of whether conditions are getting better or worse and what areas of the economy need attention.

Similar key measures could be used to monitor the state of education and other issues in the nation. In 2010, a Key National Indicator System (KNIS) was signed into law (see P.L. 111-148; H.R. 3590-562), and work to prepare for full-scale implementation is ongoing. The system is to be overseen by the Commission on Key National Indicators, an eight-member body with two members appointed by each of the majority and minority leaders of the U.S. House of Representatives and the U.S. Senate. The work to implement the system is to be carried out by the National Academy of Sciences (NAS). A total of $70 million in public financial support is authorized for a KNIS over nine year. The system is intended to "deepen our factual knowledge and understanding of the country's most pressing issues" pertaining to the economy, the environment, and people (including families, health, education, civic engagement, and culture).[1]

As part of that effort, the NAS held a workshop in January 2012, to explore possibilities for a set of key indicators that will help policy makers and the public assess the state of education in the country.[2] The broad goals for the national indicator system include providing a means for the nation to use a "shared set of facts" in determining "where we've been, where, we are, and whether we are leaving the country a better place for future generations."[3] The key task in developing education indicators will be to identify a clear and parsimonious set of measures and data that will be easy for nonspecialists to understand but which will also do justice to the complexities of the disparate U.S. education system. These indicators will be drawn from a large, often confusing and sometimes conflicting, body of information about students, teachers, schools, districts, and states.

---

[1]See http://www.stateoftheusa.org/ for current information about the KNIS.
[2]Initial choices about measures to be included in a Key National Indicators System will be made by consensus panels convened by the NAS. The process will include opportunities for public comment.
[3]See http://www.stateoftheusa.org/.

The Steering Committee on Key National Education Indicators was charged with organizing a workshop focused on exploring potential indicators that would reflect current research and address the interests of practitioners, policy makers, parents, and the general public. It was asked to commission one or more experts to develop prospective frameworks that could guide the development and implementation of a set of key education indicators and to identify candidate lists of indicators. These indicators might relate to social, economic, and other determinants of education outcomes, as well as outcomes in other sectors that are in turn affected by education. The steering committee was not asked to oversee the formal selection of a list of key indicators, or to come to any consensus about which were most promising, but rather to explore the possibilities and the primary issues to consider: the formal statement of task is shown in Box 1-1. The steering committee's role was limited to planning the workshop, and this report has been prepared by a rapporteur as a factual summary of what occurred..

In carrying out this charge, the steering committee reviewed available information about other efforts to report on education indicators, including:

- The Composite Learning Index produced by the Canadian Council on Learning (see http://www.cli-ica.ca/en/about/about-cli/what.aspx);
- The *Condition of Education* reports produced annually by the National Center for Education Statistics (see http://nces.ed.gov/programs/coe/);
- The *Education at a Glance* reports produced annually by OECD (see http://www.oecd.org/document/2/0,3746,en_2649_39263238_48634114_1_1_1_1 ,00.html);
- *Education Counts: An Indicator System to Monitor the Nation's Educational Health* (Special Study Panel on Education Indicators, 1991);
- The *European Lifelong Learning Index* produced by UNESCO (Hoskins, Cartwright, and Schoof, 2010);
- *Indicator Systems for Monitoring Mathematics and Science Education* (Shavelson, McDonnell, Oakes, Carey and Pikus, 1987);
- The *Kids Count* data book reports produced annually by the Annie E. Casey Foundation (see http://www.aecf.org/MajorInitiatives/KIDSCOUNT.aspx); and
- The *Measuring Up* reports produced biennially by the National Center for Public Policy and Higher Education (see http://measuringup2008.highereducation.org/).

They also reviewed the guidance about education indicators offered by Blank (1993), Bradburn and Fuqua (2010), Bryk and Hermanson (1993), Elliott (2009), and the U.S. Government Accountability Office (2011).

This review highlighted the plethora of education indicators that are currently available. The steering committee decided that the most effective use of the workshop would be to explore ways to select the most important ones among all the available indicators. To structure the workshop, the steering committee developed a framework (described more fully later in this chapter) that covered the stages of life in which education occurs, and commissioned a set of researchers who would each identify and justify a candidate list of three to five indicators in their own area. The workshop was then an opportunity for a diverse group of researchers and policy makers to consider two questions:

1. Given the state of the field, are there key education indicators so widely recognized that they would form a natural core of the education component of the KNIS?
2. What additional work is needed to support a formal recommendation of education indicators to be included in the KNIS?

This report describes the workshop presentations and discussions and is intended as the first step in the development of a portfolio of key national indicators of progress in education.

DEFINING INDICATORS

The short answer to the question "What is an indicator?" is that it is a measure used to track progress toward objectives or to monitor the health of an economic, environmental, social, or cultural condition over time. Different sorts of measures are used in different contexts. For example, the unemployment rate, infant mortality rates, and air quality indexes are all indicators. In the field of education, school districts typically collect average scores on a standardized reading assessment for each grade to monitor how well students are meeting basic benchmarks as they progress in reading. Other commonly used education indicators include high school graduation rates, rates of truancy, ratios of teachers to students, and per-pupil expenditures, as well as measures of less quantifiable factors, such as teachers' and students' attitudes.

Indicators—literally signals of the state of whatever is being measured—can cover outcomes, the presence or state of particular conditions, or the effectiveness of management approaches (National Research Council, 2011a). They can be used to measure change over time or for comparisons among outcomes, conditions, or measures of effectiveness in different places. Although indicators are usually quantitative, they may be either straightforward measures of a single phenomenon, such as the number or percentage of students who graduate in a given year, or composite measures. A composite indicator is a measure of a more complex phenomenon, such as college readiness, and may incorporate a number of variables that capture aspects of what is being measured. Thus, an indicator is not the same thing as a statistic. As a primer on education indicators explained, statistics "need context, purpose, and meaning if they are going to be considered" indicators (Planty and Carlson, 2010).

A system of indicators, a recent report from the U.S. Government Accountability Office (GAO) notes, is "an organized effort to assemble and disseminate a group of indicators that together tell a story" about a jurisdiction or a particular issue (U.S. Government Accountability Office, 2011, p. 57). A comprehensive key indicator system—such as the KNIS—is designed to collect only a limited number of the most important indicators on a wide range of economic, environmental, and social and cultural issues of interest in the country. It is not intended to provide a comprehensive and in-depth database on specific issues.

The selection of a short list of indicators for education, to be part of a comprehensive indicator system, poses challenges both because these few indicators will be expected to distill a complex set of issues into a concise story and because they are intended to assist policy makers and guide action. A brief look at the history of education indicators highlights some of the challenges and the state of the field as preparation for a national system of indicators gets under way.

CONTEXT

Interest in developing a system of education indicators dates at least to the 19th century. An 1867 federal law called for a department of education "for the purpose of collecting such statistics and facts as shall show the condition and progress of education in the several States and Territories" (quoted in Planty and Carlson, 2010, p. 9). Interest in this endeavor has varied since then, however. The early 20th century was a time of

4

blossoming interest in the use of data in the social sciences, and the Russell Sage Foundation, an early leader in social science research, published numerous reports on education and other topics in that period, including one that ranked states according to such indicators as school attendance and school expenditures (Russell Sage Foundation, 1912). By the middle of the century, a second wave of interest in education data was evident, and reports from the United States Department of Health, Education, and Welfare (HEW), the Bureau of the Census, and others were providing substantial amounts information about the educational status of the country (Bradburn and Fuqua, 2010).

For example, HEW published *Toward a Social Report* (1969), which charted social progress using indicators covering learning, science, and art; health and illness; social mobility; the physical environment; income and poverty; public order and safety; and civic participation and alienation (Elliott, 2009). The U.S. Census Bureau (1976) produced *STATUS*, a monthly chart book covering social and economic trends. Its fourth edition featured a special report on education, and included such indicators as participation in education at all levels, expenditures on education, educational disparities, programs for students with special needs, achievement, and public views about education; it also provided cross-national comparisons. The National Education Association produced state comparisons for indicators related to the teaching profession, and the United States Department of Education's Center for Education Statistics published a variety of statistics (Ginsburg, Noell, and Plisko, 1988).

These reports, though valuable, were not part of a sustained effort to monitor key aspects of public education, or to measure student achievement (Ginsburg, Noell, and Plisko, 1988). The publication of *A Nation at Risk* in 1983 is widely credited with stimulating an earnest and sustained push for close monitoring of the system (Bryk and Hermanson, 1993; Ginsburg, Noell, and Plisko, 1988). In 1984, the Department of Education produced a 1-page summary of statistics (the "wall chart"), which made rough comparisons among the states on a set of input and outcome characteristics related to educational performance. The limitations of the data available for the wall chart spurred interest in improving the validity of state-by-state comparisons of student achievement (Ginsburg, Noell, and Plisko, 1988), and soon afterwards there was a push for increased sampling for the National Assessment of Educational Progress (NAEP) that would make it possible to report state-level achievement data.

Interest in education indicators has continued to grow. As one observer noted two decades ago, "Hardly an educational agency or group at the national or state level has *not* been involved in the business of education indicators (Smith, 1988, p. 487). In 1984, the National Center for Education Statistics began using indicators in its regular report, *Condition of Education* (Bradburn and Fuqua, 2010; Smith, 1988). The National Science Foundation stimulated a number of new initiatives, including funding a 1988 report that proposed a framework for monitoring mathematics and science education that addressed teacher quality, course content, and student achievement, and other factors (National Research Council, 1985; see also National Research Council, 1985, 1988). Then-President George H.W. Bush's America 2000 plan emphasized accountability, monitoring, and data collection for six national goals for education. State data in these six goal areas became available in 1991 when the National Education Goals Panel produced its first reports.

During the same period, the Hawkins-Stafford Elementary and Secondary School Improvement Amendments of 1988 (P.L. 100-297) authorized the establishment of a Special Study Panel on Education Indicators. This panel was chartered by the Department of Education in 1989 and produced *Education Counts* in 1991 (Special Study Panel on Education Indicators, 1991). The report lays out a conceptual framework for an ongoing indicator system that would track enduring educational issues; it identifies six critical issue areas that such an indicator system should address:

1. learner outcomes;
2. quality of educational institutions;
3. readiness for school;
4. societal support for learning;
5. education and economic productivity; and
6. equity (measures of resources, demographics, and students at risk).

The report has served as a guiding framework for decisions about the content and format of the *Condition of Education* reports issued annually by the National Center for Education Statistics (NCES). Of particular importance was the report's focus on monitoring the outcomes of education rather than on complex causal factors (John Ralph, Program Director, National Center for Education Statistics, personal communication, 2012). At present, the National Center for Education Statistics, of the United States Department of Education, and many other organizations collect and publish data on education, much of it in the form of indicators in particular areas. The *Condition of Education* report documents various sorts of data to provide detailed information in five areas: participation in education, learner outcomes, student effort and educational progress, and contexts of elementary and secondary education as well as higher education[4] Editorial Projects in Education, the publisher of *Education Week* and the *Education Counts* reports, has also published indicators in many areas, as have other organizations.

International organizations have also focused on indicators. The European Lifelong Learning Indicators (ELLI), developed by Bertelsmann Stiftung (a private educational foundation) based on domains defined by the United Nation's Educational, Scientific, and Cultural Organization (UNESCO) (Hoskins, Cartwright, and Schoof, 2010). These indicators are widely recognized for pushing thinking about ways to measure types of learning that have not traditionally been included in formal measures, though they also address academic learning (the ELLI indicators are discussed further in Chapter 6). The Programme for international Student Assessment (PISA) and the Trends in International Mathematics and Science Study (TIMSS) are other sources of international education data.

The push to establish KNIS began in the early 2000s, with a request from Congress for the GAO to explore the feasibility of a system of national indicators that could chart the progress of the country in major policy areas (U.S. Government Accountability Office, 2004). This effort led to the passage of the law authorizing the KNIS. The NAS has begun preliminary work with SUSA to prepare for implementation

---

[4]For details, see http://nces.ed.gov/programs/coe/.

by considering possible indicators in health (see Institute of Medicine, 2009) and education (beginning with this project).

PROJECT GOALS

The steering committee wanted to be sure that the complete trajectory of education across the life span would be covered in the discussion, and also that different aspects of education would be addressed. While acknowledging that other frameworks could serve a similar purpose, it developed a rough framework based on those two goals: see Table 1-1. The framework covers the stages of life by identifying five broad sectors of education: preschool, K-12 education, higher education, other postsecondary education and training, and lifelong or informal learning (learning that occurs outside the formal structures of the education system). It also identifies three aspects of education: institutions, service providers, and resources; individual-level behaviors, engagement, and outcomes; and contextual factors that influence learning. The committee emphasized that this approach is just one way to structure the discussion, and that other frameworks may be equally or more useful.

For the workshop, the committee commissioned researchers who focus on education in each of these five general phases of life to think about what information will be most needed to measure the state of education in the country as it adapts to fast-changing technologies and global economic forces. The committee asked the researchers to make presentations in which they addressed the following questions:

1. In your opinion, what are the three key indicators about (a) institutions, service providers, and resources; (b) individual level behaviors, engagement, and outcomes; and (c) contextual issues for [the specific life stage]?
2. What is the evidence base that justifies the use of these indicators? That is, what is the evidence that they matter? What is the argument for using them?
3. In what direction would we want the indicator to change over time? That is, please talk about whether we would want to see the indicator increase, decrease, or stay the same and how such changes would be interpreted.
4. What are the potential consequences associated with these indicators? That is, if people begin paying attention to them, what consequences (intended and unintended, positive and negative) may result?
5. What are the equity issues to consider for these indicators?
6. What are the relevant data sources for these indicators?

The steering committee also commissioned a set of experts to assist in synthesizing the information presented by the various panelists. This group was asked to reflect on the common themes that emerged from the discussion, paying particular attention to areas of agreement and disagreement. They were asked to address the following questions:

7

TABLE 1-1 Framework for Education Indicators Developed to Guide the Workshop

Stages of Education/Learning	Indicators About Institutions, Service Providers, and Resources	Indicators About Individual-Level Behaviors, Engagement, and Outcomes	Indicators About Contextual Factors
Birth to age 5			
K-12			
Higher education			
Other forms of postsecondary education and training			
Lifelong, informal learning			

1. How do you envision that a national system of education indicators would be used? In what ways might these indicators support policy?
2. Based on the workshop discussions, what do you think are the most important variables/statistics to include in a national system of education indicators? What are the reasons for including these indicators?
3. What are the most important gaps that have emerged between currently available data sources and the types of data that would be needed to support a national system of education indicators? What additional data are needed?

The panelists who suggested indicators were encouraged to strike a balance between ideal and realistic goals. That is, they were instructed to be forward looking in thinking about measures that will assess both current conditions and trends as well as anticipated future conditions, and not to feel constrained by what has been done in the past or their knowledge of practical obstacles to collecting particular types of data. At the same time, they were asked to consider the extent to which data would be available to support the indicators they suggested. A rule of thumb suggested by David Breneman, the steering committee chair, was that a good system might begin with approximately one-third indicators based on existing data sources with long-term trend lines, one-third indicators that are not routinely collected but conceivably could be collected, and one-third indicators in a more gray area—those that reflect areas of key importance but present measurement challenges and require further development. As he noted at the beginning of the workshop, "We have some freedom to think in a fairly open way about what we would like to see." Identifying a short list of indicators that "people should attend to and making them readily available on a website would be a great contribution to education in this country," he added. All of the workshop participants were asked to consider several questions as they considered the indicators that were discussed over the course of the two days:

1. Is the indicator from a reliable source?
2. Is the indicator reasonably accurate, precise, reliable, valid, and unbiased?
3. Is it available over time and measured in a consistent way over time? Will it continue to be available?
4. Does it reflect a salient outcome or measure of well-being?
5. Does the indicator have a relatively unambiguous interpretation? Will it be easily understood by the public?
6. Can the indicator be disaggregated in order to report it subnationally, for various population groups, and by specific demographic characteristics?

Together, the participants brought a wide range of ideas to the workshop, suggesting a range of phenomena that might be measured and a range of approaches to data collection and data systems.

Many factors should influence the selection of a short list of indicators for this broad national purpose. As Chris Hoenig, senior advisor to the NAS presidents, noted in opening remarks, the overall set of indicators ultimately adopted by SUSA will be vitally important because they will be used to guide goals and decisions about each major sector in the country. He showed the group the preliminary version of the interactive website, which displays the health indicators that have been selected, to illustrate how useful the program can be. He stressed the importance of the logical framework underlying the selected indicators, which will guide thinking about forces that shape outcomes as well as disparities.

Diana Pullin reinforced this point in her opening remarks, noting that the indicators that are ultimately selected will reflect a particular conception of what it means to be an educated person. There is sometimes a tension in discussion of public education, she added, between the goals of providing a public benefit (a populace that is equipped for citizenship and work, for example) and providing a benefit to individuals (the intellectual tools to pursue a fulfilling life, for example). There is a risk that summative data about this complex enterprise may tend to commodify it, as some have suggested has occurred with indicators used in published report rankings of colleges and high schools do.[5] Indicators can be used not just to indicate what has happened, she noted, with reference to the 2011 GAO report, but also to promote progress, to provide transparency, to further accountability, to promote civic engagement, to further economic productivity, and to engender conversation in communities and in commerce.

Like the workshop, this report is organized by the five life spans. Chapters 2 through 6 describe the indicators proposed for the stages and the issues each presented. All of the indicators suggested for each stage are listed at the beginning of these chapters in a table, so that it will be easy to see the range of what was proposed and any areas of overlap.[6] The suggestions made by the panelists are summarized, and then the key issues that emerged in discussion are described. Chapter 7 summarizes the remarks of the synthesis panel and the ensuing discussion (the workshop agenda appears in Appendix A and the particpants are listed in Appendix B). The discussion was wide-ranging and this report was designed to capture the most important themes and issues.

---

[5]See http://www.usnews.com/rankings for more information on these rankings.
[6]The presenters took varying approaches to the assigned task and this summary report reflects that. Thus, some chapters contain more references and quantitative information than others.

2
Indicators for Preschool

The period from birth to age 5 is critical to children's development and to their success in school. As Ana Gutierrez noted, "if we don't get preschool right we're not going to get the rest of it right." Disparities in the cognitive and social skills necessary for school that are evident by the time children reach kindergarten are likely to affect their progress throughout their schooling, and often intensify (Haskins and Rouse, 2005). Decisions to fund preschool in many states are among the evidence that the importance of early education has become better recognized, but gaps in access to and the quality of early education persist (Magnuson, et al., 2004).[7]

One reason for the gaps is that the early childhood sector is complex (National Research Council, 2011b). Providers of care and education vary significantly across many dimensions, including how they are funded, the preparation and qualifications of their staffs, and the nature and quality of the experiences they provide to children and families. Providers are regulated by the states so there is also variety in the rules and requirements that govern them, though national organizations such as the National Association for the Education of Young Children have worked to develop uniform guidelines for professional practice.

Every state now has a council focused on developing data systems to cover early childhood, noted Steve Barnett, but this work is just beginning. These councils are "trying to figure out what data they should collect, how they should collect it, and how to bring together information across multiple government agencies into a single system," he added, so there is still time to influence them. A national indicator system that addresses preschool would have an important influence, not just on state data collection, but policy in general, he and others agreed. All of the indicators suggested for the preschool stage are listed in Table 2-1. The sections below describe the presenters' suggestions in greater detail.

---

[7]The steering committee recognized that parents and the home environment play a key role in children's development. However, the task was to focus on education, assuming that the needs of families would be addressed elsewhere in the indicator system, though some contextutal factors were addressed at the workshop.

TABLE 2-1 Indicators Suggested for the Preschool Stage

CHARACTERISTICS OF INSTITUTIONS, SERVICE PROVIDERS, AND RESOURCES
Use and availability of care and education outside the home, including: <ul><li>Percentage of young children receiving early care and education outside the home by age 3 or 4</li><li>Type of care used by preschoolers, and the amount of it they have per week</li><li>Percentage of children in two age ranges (0 to 3 and 3 to 5) receiving different types of early care (center-based, child-care home, or informal)</li></ul> Index of the quality of care and education programs. Possibilities include <ul><li>Spending per child at each age (year of life prior to kindergarten entry), nationally and at the state level</li><li>Child-to-staff ratio or teacher characteristics (qualifications)</li><li>Direct measure, through observation, of the environment and practices in programs</li><li>Programs in which interactions are warm and stimulating, curricula have scope and sequence, children's progress is monitored, and staff receive intensive coaching</li><li>Percentage of parents whose early childhood care and education providers interact with them in a productive way</li><li>Percentage of childcare and education settings (by type) that provide emotionally supportive, cognitively stimulating care</li></ul>
INDIVIDUAL OUTCOMES
Indicator of children's language (not only in English), academic, attention, and social skills as they enter school and possibly at age 3—a NAEP-like assessment could be used
CONTEXT
Measures of the home environment and early experiences. Possibilities include <ul><li>number of parents who demonstrate (or number of children who experience) responsive, sensitive interactions</li><li>children who experience secure attachments with caregivers</li><li>percent of families that provide (or percent of children who receive) enriching and stimulating home environments</li><li>percent of families with multiple risk factors</li><li>number of children or families receiving health, mental health, and social services</li></ul> Index of children's prenatal exposure to hazards Family demographics

CHARACTERISTICS OF INSTITUTIONS, SERVICE PROVIDERS, AND RESOURCES[8]

Use and Availability of Early Care and Education Outside the Home

A starting point for tracking early education is to have a clear measure of how many and which children are enrolled in programs and the nature of those programs. Parents are the strongest influences on young children's learning and development (NICHD Early Child Care Research Network, 2002), Barnett explained, but it is clear that high quality preschool education can have substantive and lasting impacts on learning and development (Barnett, 2011; Camilli et al., 2010; and Vandell et al., 2010).[9] He advocated tracking *the percentage of young children receiving early care and education outside the home by age 3 or 4.*

The most commonly cited data on program participation are from the Current Population Survey, the National Household Education Surveys program (NHES), and the decennial census (American Community Survey), Barnett noted. These sources provide very different estimates, perhaps because of differences in how and when they ask questions about enrollment, as well as differences in definitions of age spans and program types. A single, accurate count is badly needed, in Barnett's view, and it should be disaggregated by age, income, ethnicity, geography and, if possible, type of program. There are large differences in participation by child and family background characteristics as well as location. There are also important differences between center- and home-based programs, and there is wide variation in quality (see National Research Council, 2011b for discussion of this issue).

Barnett focused on the number of children receiving any type of care, but Margaret Burchinal suggested that it is also important to measure the type and number of institutions, service providers, and resources that are available, using the data sources Barnett had mentioned. She suggested documenting *the type of care used by preschoolers and the amount of it they have per week* (NICHD Early Child Care Research Network, 2002). Similarly, Deborah Vandell suggested tracking *the percentage of children in two age ranges, 0 to 3 and 3 to 5, by the different types of care they receive, center based, home based, or informal.*[10] She noted that these three basic types capture some of the extreme range in quality of child care and education. She also noted that center-based care, such as Head Start or formal pre-K programs, has been shown to benefit children in terms of cognitive development and academic performance at entry to preschool. Low-income children are less likely to attend high quality programs, she

---

[8]The discussion of the indicators in this and other chapters follows the broad structure of the steering committee's framework, but the indicators suggested for each stage fell naturally into different arrangements. When possible, similar indicators are discussed together under a general heading, with specific ones highlighted with italic text.

[9]The effects are generally larger for cognitive than for social and emotional domains, Barnett noted, but the latter may be more difficult to measure effectively.

[10]Researchers distinguish among (1) center-based child care, which is provided in a facility that may or not be part of a larger institution; (2) home-based childcare, in which a paid provider cares for one or more children in a private home (generally distinguished from arrangements where babysitters or nannies work in the child's own home); and (3) kith and kin care, which is care provided by relatives or friends, who may or may not be paid. Kith and kin care is not subject to state licensing standards.

added, and it is important to have a clear picture of variations in quality and access. In addition to NHES, she noted, the Survey of Income and Program Participation could be used for this purpose.

Quality of Care and Education Programs

The quality of programs is an essential determinant of their effectiveness, Barnett argued, and access to quality programs varies by age, family background, and geography (Barnett et al., 2010; Camilli et al., 2010; Karoly et al., 2008; Pianta et al., 2009). Parents do not have an easy time distinguishing quality, he noted, so even high-income children may be enrolled in "pretty poor quality places," he added.

Quality is complex, however, and current measures are far from satisfactory, he added (see Burchinal, Kainz, and Cai, 2011). The National Institute for Early Education Research publishes an annual report that includes data on spending for child care and education, teacher-child ratios, and other indicators, but in Barnett's view, these data do not provide a picture of all preschool programs.

A composite index is likely to be most useful for this purpose, in Barnett's view, because no one measure would capture the important aspects of quality. He suggested a few possibilities: *spending per child at each age (year of life prior to kindergarten entry), nationally and at the state level; child-to-staff ratio or teacher characteristics (qualifications); and direct measure, through observation, of the environment and practices in programs.* Direct observation is complex and expensive, he acknowledged, but it probably offers the strongest potential for capturing the key components of quality.

Burchinal also focused on the quality of the programs and providers that are available, arguing that this information is essential to addressing the achievement gaps among groups of children. She observed that such simple indicators as *teacher-child ratios and teacher credentials* show only very modest relationships with outcomes for children, so even though they are the easiest to use, they are "not really great indicators" (Burchinal, Kainz, and Cai, 2011). She focused instead on what has been shown to affect outcomes and suggested measuring the extent to which:

- *caregivers provide frequent interactions with all children that are warm, responsive, and linguistically rich* (see, e.g., Burchinal et al., 2010; Gormley et al., 2005; Grindal et al., 2011; Peisner-Feinberg et al., 2001; Shager et al., 2011;Vandell et al., 2010);
- *programs implement focused curricula with scope and sequence and monitor children's progress* (see Clements and Sarama, 2008; Fantuzzo, Gadsden, and McDermott, 2010; Bierman, Nix, Greenberg, Blair, and Domitrovich, 2008);
- *programs identify children who are not making adequate progress and provide interventions for them* (Buysse and Peisner-Feinburg, 2010; Fox, Carta, Strain, Dunlap, and Hemmeter, 2010); and
- *programs use intensive coaching linked to their curriculum or to promote high-quality teacher-child interactions* (Pianta, Mashburn, Downer, Hamre, and Justice 2007; Powell, Diamond, Burchinal, and Koehler, 2010; Vernon-Feagans, Kainz, Amendum, Ginsberg, Wood, and Bock, 2012).

Many early care and education systems[11] do monitor some or all of these features, Burchinal pointed out. For example, the Head Start program uses the Classroom Assessment Scoring System (CLASS). Also useful, she suggested, would be Head Start's Family and Child Experiences Survey (FACES) and the Performance Information Report (PIR) that is conducted in every program. About two-thirds of states have quality rating and improvement systems, Burchinal added, and more than half include an observational rating system designed to measure caregiver responsiveness and program stimulation (Tout et al., 2010). Many state pre-K programs also include an evaluation that measures quality using an observational rating system, and many states also have quality rating and improvement systems for child care and education programs. She also suggested that a NAEP-based program could be useful in the preschool context, as discussed below.

A related indicator suggested by Sue Sheridan was *the percentage of parents whose early childhood care and education providers interact with them in a way that is planned and collaborative and supports the development of learning goals and the successful navigation of the transitions into kindergarten and elementary school.* She noted that children who experience discontinuity between home and school are at risk for decreased academic performance, and that, conversely, positive partnerships between professionals and families produce positive changes in the family environment, parent-child relationship, parenting skills, and family involvement in children's learning (Caspe and Lopez, 2006; Grolnick and Slowiaczek, 1994; Masten and Coatsworth, 1998; Phelan, Davidson, and Yu, 1998; Sheridan, Bovaird, Glover, Garbacz, Witte, and Kwon, in press; Turnbull et al., 2011). Positive interactions also develop patterns and influence parenting and developmental trajectories for children showing early signs of educational risk, she added (Jung 2010; Turnbull et al., 1999).

Vandell approached the issue of quality by proposing a measure of *the percentage of childcare and education settings (by type) that provide emotionally supportive and cognitively stimulating care.* She noted that data from the Early Childhood Environment Rating Scale (ECERS) and the preschool section of the Classroom Assessment Scoring System could be used for this indicator, and that other measures in development might also be used (Pianta, Hamre, and Laparo, 2008). She also questioned Barnett's concern that observations would be prohibitively expensive, noting that there are reliable and valid measures that could be administered with a nationally representative sample. She highlighted the importance of finding a way to consistently measure quality, commenting that "for us to have [no] idea about the quality of care for the first five years of life in the U.S. is appalling."

---

[11]Researchers and others often use the phrase "early childhood care and education" to capture the idea that anyone working with young children will be providing care and that purposeful education of young children (beyond what any caring adult might offer) is an additional element that both child care centers and preschool programs offer.

INDIVIDUAL OUTCOMES

Index of Children's Learning and Development

The first 5 years of life are a time of rapid learning and development that is critical for many later outcomes, Barnett noted. Monitoring of this development is important because large ability gaps[12] become apparent well before children enter kindergarten (Halle et al., 2009; Lee and Burkham, 2002). Differences in cognitive abilities between black and white children are evident before age 3 (Burchinal et al., 2011; Dickenson, 2011). Early differences in abilities increase through age 5 and persist at the same or larger levels thereafter. Cognitive abilities, including language, attention, and math skills are strongly predictive of achievement at the end of primary school and even into high school (Dickinson, 2011; Duncan et al, 2007; National Mathematics Advisory Panel, 2008; Verhoeven, van, Leeuwe and Vermeer, 2011).

Ideally, Barnett suggested, *children's development would be measured at age 3 and at kindergarten entry.* He noted, however, that there are few comparable instruments that could be used to measure development at these stages nationally, though data from the ECLS-B could provide a start. "It's very hard to do good assessments of kids before they are 3 and pretty darn hard at 3," he observed. It's also much less expensive to collect data on 5-year-olds, he added, "but in some ways it's a little late." Nevertheless, he suggested that a NAEP-type assessment for children entering kindergarten could possibly be developed for this purpose.

Burchinal agreed that a NAEP structure is a possible tool for assessing children's readiness for school (she also suggested it could be useful for program evaluation). She noted that Head Start and public pre-kindergarten programs already have standards for developmental milestones, and often collect data to determine whether the standards are being met, so there may be other ways to accomplish the goal of establishing an indicator of *children's acquisition of the language, academic, attention, and social skills they will need in school.* There is strong evidence, she explained, that these skills predict their academic and social trajectories well (see, e.g., Alexander and Entwistle, 1988). The Head Start and state preschool data discussed could also be used for this indicator.

The advantage to measuring school readiness using NAEP, in her view, is that such a program could monitor in a uniform way the children and programs in all early childhood systems or contexts. It might be possible to combine observational measures with other measures to improve the estimation of quality, as has been advocated in a K-12 context.[13] She was more optimistic than Barnett about possibilities for assessing young children, noting that language skills, for example, are relatively easy to measure and are quite predictive of later development.

Vandell also advocated the use of *a NAEP-like assessment of children's reading and mathematics skills and their approach to learning at the beginning of kindergarten.* For her, the key features would be collecting the data through a direct assessment and using a nationally representative sample. The Early Childhood Longitudinal Study

---

[12]Barnett commented that he prefers to use the term "gradients" because there is such a clear, continuous and linear relationship between income and test scores, as well as social and emotional development.

[13]For more information, see http://www.metproject.org/.

(ECLS-K), which already collects data from a nationally representative sample of young children, could possibly be used for this purpose, she added.

CONTEXT

Measures of the Home Environment and Early Experiences

Early family experiences have large effects on children's development and are highly predictive of long-term development, and this issue was thoroughly addressed by the presenters. Most agreed that measures of these factors are important, but they acknowledged that obtaining accurate and reliable information can be challenging, given privacy concerns raised when data are collected about families and community environments.

Developmental differences emerge early, and many are associated with complex family circumstances that vary within and across ethnic groups, Barnett explained. Indeed, much of the variation in preschool abilities is explained by differences in early parenting and to a lesser, but still important, extent by differences in experiences with other caregivers and preschool education. For example, variations in children's very early experiences with language—how many words are spoken to them, how often they are spoken to, the extent to which they are encouraged to use their own developing language abilities and emergent pre-literacy skills—predict not just their early vocabulary, but also their vocabulary in elementary school (Dickinson, 2011). A similar relationship is found between early home mathematics experiences and early knowledge of mathematics (Levine et al., 2011).

Some surveys, such as the National Household Education Surveys Program, already collect information about aspects of the home environment, Barnett explained, but constructing an index of the home environment and parental behavior that would be informative for policy makers and in communicating with the public, and yet not so simple as to be misleading, will take some thinking, in his view. For example, if one measure is number of books in the home, simply buying books for families could be the logical policy response, yet it would be unlikely to have much effect.

Sue Sheridan agreed, and she focused most of her suggested indicators on aspects of the home environment. Her first suggestion was *the number of parents who demonstrate (or number of children who experience) responsive, sensitive interactions, such as displays of affection, physical proximity, and positive reinforcement.* She explained that the evidence base for the importance of parenting is strong, noting several specific findings:

- Sensitive, responsive parent-child interactions and relationships are associated with stronger cognitive abilities, school achievement, literacy skills, and social competence (Auerbach, 1989; Brice-Heath, 1986; NICHD Early Child Care Research Network, 2002; Snow, 1988, 1991; Tomasello and Farrar, 1986).
- Parent-child interactions that include displays of affection, physical proximity, positive reinforcement, and sensitivity are associated with children's positive cognitive growth over time (Bradley et al., 2001; Burchinal et al., 1997; Landry et al., 2001; Pungello et al., 2009; Rao et al., 2010).

16

- Preschool children who experience secure attachments with a caregiver as infants have been shown to engage in more spontaneous reading activities and perform better on literacy measures, in comparison to those who were insecurely attached (Bus and van IJzendoorn, 1988). Those with secure attachments demonstrate greater levels of curiosity and self-directedness, are sensitive to others' needs, and are eager to learn (Bost et al., 1998). They also experience higher levels of behavioral and emotional control, exhibit greater adaptability and autonomy, and experience fewer difficulties approaching learning tasks (Sroufe, 1983).

Sheridan also suggested measuring *the percentage of families that provide (or the percentage of children who receive) enriching and stimulating home environments.* Such environments, she explained, are characterized by rich and responsive verbal exchanges (reciprocal turn-taking and open questions); shared interactive experiences with books and other printed materials; exposure to vocabulary enrichment; positive social interactions; guided exploration and constructive play; and opportunities for learning and problem solving.

Parental engagement behaviors and a stimulating home environment in children's first 5 years are highly related to cognitive and social outcomes, she explained, and play an exceptionally important role in shaping the capacity of the developing brain. Research has demonstrated, she added, that a cognitively, emotionally, and physically stimulating environment is related to high language and cognitive skills, school readiness, and academic success at least through the primary grades (e.g., Arnold et al., 1994; Bradley, Burchinal, and Casey, 2001; Chazan-Cohen et al., 2009; Espinosa, 2002; Foster et al., 2005; Hill, 2001; Hood, Conlon, and Andrews, 2008; NICHD Early Child Care Research Network, 2002; Pan et al., 2005; Raikes et al., 2006; Tamis-LeMonda and Bornstein, 2002; Weigel, Martin, and Bennett, 2006a, 2006b).

The frequency and quality of words a child hears during his or her first 3 years of life are critically important in shaping language development, she added, and interventions to support parenting skills and learning have been shown to improve children's outcomes in a number of areas (Hart and Risley, 1995; Knoche et al., in press; Landry et al., 2001; Sheridan et al., 2010, 2011).

Sheridan also addressed risk factors in the home, suggesting as an indicator *the percentage of families with multiple risk factors, such as poverty, single adult household, non-English-speaking household, parents with less than a high school education, a teen parent, and parental mental illness.* She explained that young children's brains are extremely vulnerable to the effects of adversity, and that deprivation, neglect and the other factors she had listed are harmful in themselves; when multiple risk factors are present, the probability that children's development will be compromised increases (Bornstein and Bradley, 2003; Brooks-Gunn and Markman, 2005; Duncan and Brooks-Gunn, 2000; Essex et al., 2001; Grossman et al., 2003; Hart and Risley, 1995; National Task Force on Early Childhood Education for Hispanics, 2007; Raviv, Kessenich, and Morrison, 2004; Roberts et al., 1999; Sameroff et al., 1987, Wood, 2003).

In response to several questions from participants who objected to the inclusion of language on this list, she emphasized that she included non-English-speaking households not because speaking another language is in any way a disadvantage, but because families in which the adults do not speak English may have greater difficulty establishing connections to the public school system and other community resources and supports. This is of particular importance in areas where the community resources and supports are offered only in English, and fluency in English is needed to fully benefit from them. She stressed that this can be a disadvantage for children in families that also have risk factors such as poverty, and that her concern was with the ways in which the lack of English-speaking skills can magnify the effects of other risk factors.

A related important indicator, for Sheridan, is *the number of children or families receiving health, mental health, and social services*, because such connections buffer children against negative outcomes (Korbin and Coulton, 1997; Sampson, 1992). Partnerships with providers of these services can produce changes in the family environment, she noted, and provide potential for changing developmental trajectories, and "moving the needle, particularly for children who show early signs of developmental risk."

Index of Children's Prenatal Exposure to Hazards

Prenatal exposure to tobacco, alcohol, drugs, environmental toxins, violence, and parental stress also adversely affect child development, and the developmental consequences can be very severe and are long term. Barnett included tracking *exposure to such hazards* in his list, noting that some data are already collected: the Early Childhood Longitudinal Study Birth Cohort (ECLS-B) includes hazard exposure, and there may also be health surveys that would be useful. However, Barnett believes a new instrument would be needed to adequately capture significant exposure, though he believes collecting the data every 5 years would be sufficient because the circumstances are not likely to change quickly.

Family Demographics

For Vandell, the basic context in which early care and education are provided is defined by the demographic characteristics of families. She suggested collecting basic data on the families of young children in order to identify differences in the quality of and access to care for children in different subgroups, defined by income level, home language, and ethnic and racial background. Current data suggest that access and quality vary by group, she noted, but these family demographic factors should be collected for all developmental periods and should be disaggregated by state and by age. The Current Population Survey could be used for this purpose, she added.

ISSUES FOR THE PRESCHOOL STAGE

Four issues were the focus of most of the discussion about preschool indicators. Each of these challenges, noted Eugene García, is exacerbated by the fact that "we really don't have a system" for addressing the needs of young children. These years are every bit as important as later stages of development, and indeed set the stage for what happens in school and beyond, he added.

Assessing Young Children

Several of the presenters suggested that assessing children, possibly at ages 3 and 5, to monitor their development during the preschool years and their readiness for kindergarten (most advocated assessing only a representative sample), and participants had many thoughts about this proposal.

One approach would be to expand NAEP for this purpose, and participants noted that this could be a reasonably affordable model for sampling the national population. Adding a preschool component to NAEP would make it possible, but several practical concerns were raised. NAEP's samples are drawn from public school enrollments, but many preschoolers are enrolled in programs that are not connected to the public school system; thus, it would be necessary to identify an additional sample of children not enrolled in public pre-K programs in order to fully represent the population. NAEP is given in the spring, but if the goal is to capture kindergarten readiness, it would be necessary to assess children in the early fall. An advantage to NAEP is that it includes a background questionnaire that collects data on students, families, and schools, one participant pointed out, but privacy concerns have severely constrained what can be included on these questionnaires, to the point, in his view, that they are almost useless. Presenters had emphasized the importance of contextual information about children and families, so this could be a serious drawback to using NAEP, in this participant's opinion.

Another participant noted that sampling is a useful way to track basic trends but that if states or districts would like to have formative data that would allow them to plan instruction and to target students' needs, it would be necessary to use a different measurement approach.

Other assessment instruments might also be tapped for preschool indicators, and participants mentioned a web-based early childhood assessment used in Australia, the Performance Indicators in Primary Schools (PIPS), as well as the fact that OECD, UNESCO, and the World Bank are all interested in the development of improved indicator systems for early childhood. Using the ECLS program is another option, several people noted.

Addressing Language and Immigration

Each state has its own definition of what an English-language learner is, noted García, "so if we are going to measure it we probably ought to have some consensus about what it means." Language development is a central aspect of the first 5 years of life, and language mediates children's relationships with their teachers and caregivers. If a measure of language proficiency is concluded among the indicators, García suggested, it could have unintended consequences if it does not capture the benefits of development in the family language as well as of developing proficiency in English.

High rates of immigration to the United States mean that populations are changing, sometimes very rapidly, and cultural norms are evolving as well. Development and learning are culturally mediated, García added, and it is important to be careful about standardizing expectations for children and the adults who work with them unnecessarily. For example, expectations related to how adults communicate with children, for example, may differ significantly across groups, but a range of approaches may work equally well in helping children develop.

Capturing Quality

"We need more uniformity than we currently have in assessments of quality," noted Barnett. There are possibilities, but he believes the cost of conducting nationally representative observational studies is currently prohibitive. Other participants were more optimistic about the possibilities, but agreed that more development is needed to make measures of this critical element a reality for a national indicator system. Part of the difficulty is that while there is a fairly strong consensus about what children need—which includes secure attachments with adults; supportive and nurturing relationships with caregivers; a stimulating, language-rich environment; and systems to support those experiencing adversity—these elements are much more difficult to measure than, say, mathematical or language learning. At the same time, the children who are not getting all of those elements in a preschool program are all too often those whose families have the greatest risk factors. One participant noted that the greatest risk is in that disconnect between families and available resources.

The question of which indicators would be most likely to lead to productive change is not a simple one. Research seems to support the idea that the presence of a curriculum is a particularly important indicator of the quality of a program, for example, but the disparate nature of preschool and child care systems and their governance by states means that more work is needed to determine the best approach to measuring quality, participants suggested.

Capturing Environment and Context

The disconnect between families at risk and the supports communities have available points to a broader issue that may or not belong among the education indicators, participants suggested. The health of communities, specifically the health of infants and mothers, economic well-being, and the extent of risks such as substance use, crime and violence, and the like, have a profound influence on young children but are also important for other reasons. It may be that the health of communities along such dimensions would best be captured elsewhere in the indicator system, it was noted. So long as the elements that are critical to the development of young children are captured somewhere, they need not be included among the education indicators.

The discussion closed with a reminder that the years from birth to age 5 are at least as important as 5 to 17—and that even a simple indicator of what society is spending on the first 5 years, compared with spending on other phases of life, would be a useful way of pushing policy makers and other to reflect on its goals for supporting the nation's youngest children.

3
Indicators for K-12 Education

Schooling for students in kindergarten through grade 12 is the heart of what most people think of as "education." Public K-12 education is a large and complex system, and it is the focus of many expectations, from producing responsible and productive citizens to boosting the nation's standing in science and technology and its position with respect to its economic competitors.[14] The presenters represented an array of research experience: all the suggested indicators are listed in Table 3-1.

CHARACTERISTICS OF INSTITUTIONS, SERVICE PROVIDERS, AND RESOURCES

One way to think about the providers of public K-12 education, Mark Dynarski observed, is as an industry, and doing so highlights several aspects that are important for thinking about which indicators of quality and effectiveness would be most valuable. Public education is, by and large, provided by governmental structures that are highly decentralized, with each state having separate authority to supply education, and more than 15,000 school districts operating within those states. Thus, the supply of education, in an economic sense, takes the form of face-to-face instruction in classrooms, supported by physical capital, as economists term it, in the form of land, buildings, and equipment. Schools also are responsible for providing transportation and meals to students. Public education is supported by federal, state, and local property taxes. The federal share is less than ten percent of the cost, though the federal government exerts significant authority over education through legislation and regulation.

As with other publicly provided goods, the "output" and efficiency of public education are challenging to define and measure. National assessments of achievement, in place for decades, and, more recently, state assessments developed in compliance with the No Child Left Behind Act, provide proxy measures. But, in Dynarski's view, a set of indicators should provide a broader picture of the state of the complex system that educates nearly 50 million students in more than 98,000 schools and spends more than $500 billion per year.[15]

Other presenters had somewhat different perspectives—the group offered indicators for both schools and teachers.

---

[14]The indicators system would include data on private schools and their students but it is public education that is the primary focus of policy makers.
[15]Information avaialble at http://nces.ed.gov/fastfacts/display.asp?id=84 and http://nces.ed.gov/fastfacts/display.asp?id=84.

KEY NATIONAL EDUCATION INDICATORS

TABLE 3-1 Indicators Suggested for K-12 Education

CHARACTERISTICS OF INSTITUTIONS, SERVICE PROVIDERS, AND RESOURCES
SCHOOLS
<ul><li>Surveys of safety and orderliness of the school climate</li><li>School culture related to college and career aspirations, including :<ul><li>percentage of students who go on to two- and four-year colleges and full-time employment</li><li>surveys of students' expectations and their schools' approach to preparation for college and career</li></ul></li><li>Collaborative school community focused on student learning, using surveys of teachers, parents, and students</li><li>Surveys of parent satisfaction</li><li>High school completion rates</li><li>Grade retention rates through 8th grade</li><li>Teacher-student ratio</li></ul>
TEACHERS
<ul><li>Proportion of teachers whose evaluations distinguish them from a basic standard, using measures of their contributions to student achievement and their professional practice</li><li>Teachers with mastery-level and current knowledge of content they are teaching</li><li>Teachers with mastery-level and contemporary knowledge of child and adolescent development</li><li>Teacher-student interactions that demonstrate high levels and qualities of involvement, stimulation, and expansion of thinking and cognition, and sensitivity to students' perspectives, individual experiences, and backgrounds</li><li>Teacher-student interactions that foster relationships with and among students</li><li>Teachers providing challenging opportunities to learn in the classroom</li></ul>
INDIVIDUAL OUTCOMES
<ul><li>School attendance by age</li><li>College readiness levels by age and grade</li><li>Voter registration rate of 18- to 21-year-olds</li><li>Command of core content, using NAEP scores</li></ul>
CONTEXT
<ul><li>K-12 education spending as a share of gross domestic product (GDP)</li><li>K-12 spending per student</li><li>Percentage of K-12 education funding spent on research and development</li><li>Opportunity to learn</li></ul>

Schools

Safety and Orderliness of the School Climate

If students do not feel physically and emotionally safe at school, Elaine Allensworth argued, they will be reluctant to go to school and will have a hard time focusing on learning when they are there. Schools that are safe have higher attendance and better teacher retention; schools with safety problems tend to be those that struggle to improve learning, and have low attendance and graduation rates, high teacher mobility, and poor emotional outcomes for students. Poor discipline strategies in unsafe schools disproportionately affect minority and low-income students, she added.

Surveys, in her view, are the best way to measure the school climate. Other measures that may seem to be more objective are often biased by variation in discipline and record-keeping practices. Although she sees many possible positive outcomes if school climate is treated as an important measure, she cautioned that it is also possible that schools worried about accountability in this area might use counter-productive strategies, such as increasing suspension rates or arrests at school or increasing teacher-led instruction to increase orderliness in the classroom.

School Culture Related to College and Career Aspirations

A primary goal for schools is to prepare students for college and careers, and there could be several components to an indicator of how well they are doing this, in Allensworth's view. An indicator of *the percentage of students who go on to two- and four-year colleges and full-time employment* would provide a basic indication of school effectiveness. For more nuanced information, she suggested an indicator of *students' expectations and their schools' approach to preparation for college and career that is based on surveys*. A benefit of focusing on this indicator could be to make schools take on a greater sense of responsibility for explicitly preparing their students for the future. Doing so might also encourage collaboration across school levels, as systems consider the elements that foster preparedness even in elementary and middle school.

Collaborative School Community Focused on Student Learning

The degree to which school leaders, teachers, and parents collaborate has consistently been found to be among the strongest predictors of school improvement, school safety, and student learning gains (see, e.g., Literacy Collaborative, 2009). This important aspect of schools can be defined in different ways, Allensworth noted, but whether the focus is a sense of professional community among teachers; parent-teacher collaboration; inclusive leadership; or trust between and among teachers, parents, and principals, the research consistently shows positive effects on important outcomes. She suggested an indicator of *the nature of school communities that is measured through surveys of teachers, parents, and student.* Focusing on this measure may help schools improve coordination related to instruction, school climate, and staff stability.

23

Parent Satisfaction

Dynarski suggested an indicator of *parents' satisfaction.* The extent to which parents are satisfied with the education their children receive is an important indicator, in Dynarski's view, because it is they who are responsible for the demand for it. Their satisfaction could be viewed as a judgment about whether the funds are well spent. Their dissatisfaction could suggest either that the costs are out of proportion with the results, that their children are expressing unhappiness with school for reasons that parents judge to be valid, or that parents have incomplete or erroneous information about their schools.

The National Center for Education Statistics has measured parent satisfaction at regular intervals using the "National Household Education Survey." It asks parents about their level of satisfaction with schools, teachers, academic standards, and order and discipline in schools, and recently a question about staff interaction with parents was added. The survey results show that, since 1993, levels of parent satisfaction have been stable—a result Dynarski finds surprising considering how tumultuous the period covered has been.

High School Completion Rates

High school completion rates "can't be overlooked," noted Rob Warren. There are fewer and fewer jobs for people without high school diplomas, he noted, and failing to complete high school is a robust predictor of many economic and social difficulties, including poverty, delinquency and crime, and civic disengagement. Much recent attention has focused on the accuracy of different ways of measuring school completion and dropout rates (see National Research Council, 2011a). Measures of a few states' 4- and 6-year completion rates are available in publicly reported data (longitudinal student tracking systems), and it is possible to accurately approximate these data for past years, he observed, using counts of enrollment by grade and public school graduates (Greene, 2002; Heckman and LaFontaine, 2008).

Retention Rates Through 8th Grade

An equally important but less well-documented indicator, suggested Warren, is *grade retention.* Data on school completion have been well publicized, but few people could estimate the average grade retention rate for the nation, he noted. Grade retention is one of the leading predictors of high school dropout rates and is also a valuable indicator of students' progress through school, in terms of both cognitive and noncognitive outcomes. Interpretations of the significance of grade retention rates vary, Warren observed. To some, low rates are evidence that there is too much "social promotion," or a push to move students to the next grade even if they are not academically ready, while to others low retention rates mean that most children are succeeding. Conversely, high retention rates may look like evidence that schools are unfairly punitive or that they are holding students to high standards, depending on the observer.

To demonstrate the possibilities for exploring retention data, Warren compared data for twelve states that do report their grade retention rates with three sources of publicly available data: census-based estimates of the number of people in each jurisdiction who are of a certain age (e.g., six year-olds in Kansas in a particular year); data from the Private School Survey on the numbers of K-8 students who attend private schools; and public school enrollment counts from the Common Core of Data (CCD). Using the three data sources, he constructed estimates of grade retention rates for the 12 states and found them to be highly correlated with the rates the state had calculated. For 2004-2005, he observed a range for the 1st grade among the 12 states from less than 4 percent for Wisconsin to between 7 and 8 percent for Delaware, for an estimated rate for the United States of 4.2 percent (Louisiana was an outlier at nearly 12 percent because of Hurricane Katrina).

This method works for grades 1 through 8, he noted, but the high school grades present additional challenges.

Teacher-Student Ratio

For Dynarksi, the "fundamental technology" inside schools is the relationship between teachers and students inside the classroom. One possible indicator would be pupil-teacher ratio, a measure of the contact between teachers and students, and it is reported by NCES.

The ratio has declined steadily since 1960, when it was one teacher for every 25 students, to the current level of one teacher for every 15 students. However, Dynarksi cautioned that these figures include specialist teachers, such as reading and special education faculty who play a range of roles, so the class size likely has not declined as significantly as the ratio might suggest. For example, the average size for elementary classrooms was 21.2 students in 1999 and is currently 20.3 students. Dynarski noted that these downward trends have not yielded increases in NAEP scores in that (admittedly short) period of time. One might ask whether this finding shows that the school system has become decreasingly productive in terms of how efficiently resources are used, he added, but in considering such a question it would be critical to consider many other factors that influence the productivity of schooling—changing characteristics of student populations and changing educational goals and emphases, for example.

Teachers

Teacher Evaluations

Brian Gill suggested an indicator of *the proportion of teachers whose evaluations distinguish them from a basic standard, using measures of their contributions to student achievement and their professional practice.*

"Education researchers have come to a conclusion that has long been apparent to parents and teachers," he noted: "Teachers matter, both in the short and the long term" (see, e.g. Chetty, Friedman, and Rockoff, 2011; Hanushek and Rivkin, 2010; Kane and Staiger, 2012). This body of research has had particular policy importance as the U.S. Department of Education has focused on using its leverage to encourage states to better distinguish between high- and low-performing teachers, and to base high-stakes consequences on their evaluations. This poses both opportunities and challenges, in

Gill's view, because the policy is actually "ahead of the research." It is clear that teachers matter, he added, but there is a great deal more to learn about how they matter. Which characteristics and practices are most important to teacher effectiveness, and how they can be reliably measured are questions that have not been fully answered. Current evaluation systems, he added, are typically showing that more than 95 percent of teachers are satisfactory, so there is a clear need for better ways to distinguish among them.

Most attention has centered recently on "value-added" models that measure teachers' contributions to changes in students' standardized test results in reading and mathematics.[16] In Gill's opinion, value-added measures show promise as one component of a broader system of teacher evaluation, despite methodological challenges. The field will also need deeper and broader measures of performance that can be used with teachers of other subjects and grades not tested, and that can capture aspects of performance not measurable using standardized assessments, he added. Other measures (e.g. the Classroom Assessment Scoring System (CLASS)[17] and the Framework for Teaching[18]) have shown promise as ways to measure teacher effectiveness (Kane and Staiger, 2012).

In Gill's view, one concern about evaluation systems that hold teachers accountable for their students' gains on standardized tests is that they can create incentives for teachers to "game the system," taking steps designed to boost scores rather than to improve student learning. He believes that value-added metrics should be used with caution, that they should be used, if possible, with multiple measures of student achievement, and that they should be augmented with robust measures of teachers' professional practice.

Teacher Quality

Robert Pianta suggested two indicators of teacher quality: (1) *teachers with mastery-level and current knowledge of content they are teaching* and (2) *teachers with mastery-level and contemporary knowledge of child and adolescent development.* These two aspects of teachers (including those who interact with students in after-school programs and other settings outside the regular school day) are basic structural indicators of their effectiveness, in his view.

"Does the person teaching algebra know algebra?" is a key question, he explained. Another is whether a teacher has up-to-date knowledge of child and adolescent development, and of learning trajectories in a variety of skill domains. These would not be difficult to measure, in Pianta's view, but currently, "we do not assess them at all." The proxies most often used—possession of a master's degree, course-taking, or certification status, for example—are not associated "with much of anything," in his view. He acknowledged that a considerable amount of research has shown the importance of content knowledge, but he argued that more precise measures than teachers having majored in the subject they are teaching are needed.

---

[16]See National Research Council (2010) for more information about value added modeling.
[17]For a description, see http://www.brookespublishing.com/store/books/class/index.htm.
[18]For a description, see http://www.danielsongroup.org/article.aspx?page=frameworkforteaching.

Classroom Processes

Pianta also suggested three indicators of classroom processes: (1) *Teacher-student interactions that demonstrate high levels and qualities of involvement, stimulation, and expansion of thinking and cognition, and sensitivity to students' perspectives, individual experiences, and backgrounds;* (2) *Teacher-student interactions that foster relationships with and among students;* and (3) *the extent to which teachers provide challenging opportunities to learn in the classroom.*

Looking at teachers' characteristics is not enough, in Pianta's view. It is important also to look directly at the character of classroom instruction in three ways that he believes are reasonable and practicable. A variety of research has pointed to teacher engagement as a key factor in student learning (Gates Foundation, 2010). When teachers interact in an engaging way, he explained, they create cognitive demand through feedback and discourse, and they also help students feel a sense of belonging and appreciate the relevance of the content that is being taught. These are classrooms, he added, "in which children are trending higher" even on standardized assessments. These teacher behaviors, he suggested, also show an association with lower dropout rates and rates of social and behavioral problems (Gates Foundation, 2012).

These behaviors can be quantified and assessed, he added, using some of the observation techniques discussed in the context of early childhood education. He argued that number of large-scale studies have demonstrated that both observations and student surveys can capture the nature of classroom interactions quite well. Pianta cautioned that the rigor and reliability of such measures depend on careful design, and also that if such measures are used, it will be important to watch for unintended consequences. For example, if an indicator is adopted that "simply counts the number of open-ended questions a teacher asks, you are going to get a lot of open-ended questions," without, perhaps, much insight into whether those questions are embedded in an effective and engaging interaction.

INDIVIDUAL OUTCOMES

School Attendance by Age

A basic indicator of opportunity to learn, in Elaine Allensworth's view, is student attendance. "If kids are not in school, they are not learning," she observed, and proposed an indicator of *school attendance by age.* She emphasized that a large body of research demonstrates the importance of attendance. There is wide variation in attendance rates, even across schools serving very similar populations. Attendance is often viewed as a trivial or low-level predictor, she added, but it is highly predictive of eventual educational attainment—it influences learning, grades, and graduation rates. Attendance changes as students age, with problems typically beginning in the middle grades and rates "bottoming out" in high school. She also noted that it is relatively easy and not expensive for schools to improve attendance, and they can achieve significant benefits quickly.

Data on average daily attendance are already collected by all schools (in terms of the percentage of enrolled days that are days attended), but attendance also can be calculated to include nonenrolled students and thus be combined with information about students who drop out. Another possibility, Allensworth noted, would be to measure chronic absence, though doing so without also measuring daily attendance for all students would mean focusing on a very small subgroup of students and missing the opportunity to flag problems before they become chronic.

College Readiness Levels by Age and Grade

Allensworth proposed an indicator of *college readiness.* Although it may seem difficult to measure, colleges have long used systems based on high school grades and test scores, and grades have been found highly predictive of both college performance and later earnings. Tests scores are less predictive, she added, particularly when other factors are taken into account. Many people view high school grades as too subjective to be used as a reliable indicator, she added, but they do reflect students' engagement, motivation, and other noncognitive factors associated with academic success. Since standardized tests provide measures of certain skills, the two together provide a better picture of student progress than either does alone, in her view.

There are a variety of tests already in use, as well as databases containing information about students' coursework and high school grades, and linking those to test scores would provide the potential for developing a composite measure of readiness, Allensworth believes, and it should be possible to adapt and improve such measures to make it possible to assess students progress toward readiness in earlier grades. Though data are available at the national, state, and local levels, she added, they are not always comparable.

There is also a significant caution related to this measure, she added, because high-stakes accountability incentives based on test scores and grades can lead to poor educational practices. Emphasis on testing can lead to narrowing of the curriculum and cause teachers to devote excessive time to test-taking skills. Similarly, using grades for accountability purposes can lead to grade inflation, and, particularly in schools with very poor student engagement, grading practices may reflect effort at the expense of actual performance. These problems, she added, are most likely in schools serving the most disadvantaged students.

Voter Registration Rate of 18- to 21-Year-Olds

One of the original goals for public education, noted Gill, was to develop the educated citizenry that is essential to the functioning of democracy, and opinion polls indicate that people still believe that is an important goal. He proposed an indicator of *the voter registration rate of 18- to 21-year-olds.* This issue is even more important in the 21st century than it was in earlier in the nation's history, he added, because government plays a larger role in people's lives than in the past, and the "stream of information bombarding us" puts a high premium on the ability to distinguish fact from fiction and from opinion. New approaches to school funding and governance (e.g., vouchers and charter schools) have challenged the expectation that education will be delivered in a "common school" under the direct authority of democratically elected officials, he added. For all these reasons it is important to focus public attention on

28

preparation for citizenship and measure it not only at the national level but also for each sector (traditional public schools, charters, private schools, and home schooling), and for individual schools.

The rate at which young people register to vote and actually vote is one possible indicator of how well schools are succeeding in this mission, Gill explained, and the rate at which young people (including those too young to vote) participate in community service is another. The National Assessment of Educational Progress (NAEP) includes an assessment in civics that addresses civic knowledge, attitudes, and skills, though, Gill noted, that assessment has only been administered three times in the last 15 years.

Command of Core Content

For Rob Warren the fundamental question is whether kids are learning. He proposed an indicator of *command of core content.* He turned to NAEP scores as the best starting point for examining what students know about core subjects in multiple grades, in a way that can be disaggregated by state, by social-economic group, and across time. NAEP has some drawbacks, he acknowledged. The cut scores are sometimes seen as too high or as arbitrarily set,[19] for example, but students who lack basic proficiency in the subjects tested by NAEP are less likely to complete postsecondary schooling and are at a disadvantage in the labor market as a result. NAEP scores are already widely publicized, and the American public already views them as indicators of the effectiveness of public schools.

CONTEXT

Education Spending as a Share of Gross Domestic Product

The proportion of a country's expenditures that is devoted to education is a possible indicator of the importance the country attaches to it, and is used as a basic point of international comparison, Dynarski explained, and he proposed an indicator of *education spending as a share of gross domestic product.* The United States does not rank very high in that comparison, but that is partly because it is so wealthy, he noted: it could be spending a great deal on education without the total constituting a large share of gross domestic product (GDP). This ratio fluctuates, because the GDP fluctuates, but the trend since 1984 has been a steady increase in education spending, in real dollars.

This ratio can be calculated using GDP data and separately collected public expenditure data.

Spending per Student

It is also important, Dynarski added, to consider that if the number of students has increased significantly, expenditures per-pupil may actually decline even while the

---

[19]The cut scores for NAEP are determined through a judgment-based process in which a panel of educators and public members reach consensus about descriptions of the sorts of skill and knowledge students should demonstrate to meet performance levels. For a description of the standarrds-setting methods used for NAEP, see http://nces.ed.gov/nationsreportcard/set-achievement-lvls.asp.

proportion of GDP increases, and he suggested an indicator of *spending per student*. He noted that data collected by NCES shows that per-pupil spending has trended upwards since the early 1960s, sharply up until the 1990s and more gradually since then.

Spending on Research and Development

The expectation in many fields—particularly medicine—is that a great deal of research will be conducted and practice will follow that research as it develops, noted Brian Gill. The same has not been true in education, and the result is that "the technology in education has been remarkably unchanged for a century or more." In response to Dynarkski's observation that schools may not have become more productive, Gill suggested that stagnant productivity is partly a result of a chronic underinvestment in research and development, and he proposed an indicator of *the percentage of education funding that is spent on research and development*.

Moreover, the demands of the job market and of effective citizenship have steadily grown, making the stakes of schooling higher than they have ever been. In his view, dramatic improvements in educational productivity are needed. New technologies that have been described as revolutionary, he suggested, have not changed the core of classroom instruction. Gill believes that there is reason for cautious optimism about the potential in educational technologies that are now being developed, but that dramatic increases in productivity will require both large additional investments in research and development and a willingness on the part of school systems to allow new technologies to disrupt longstanding institutional practices

This effort should include investment in research into ways to better measure student achievement and evaluate the effects of interventions, in Gill's view. "One of the frustrations" of the new research on teachers' value added, he added, "is that it has thus far provided little information on what highly effective teachers are actually doing that makes them highly effective." Research into these and other questions is needed not only for its own sake but to support immediate needs.

Opportunity to Learn

Although it may be hard to measure, *opportunity to learn* cannot be ignored, argued Rob Warren, and he proposed it as an indicator. He likes the definition used by Grodsky and colleagues, "the resources available to students, most often in the classroom setting, that facilitate their acquisition of knowledge or skills" (2008, p. 388), and he stressed that it is important to consider this aspect of schooling separately from other indicators. Measures of achievement (e.g., NAEP scores) and measures of school completion both conflate two separate factors: (1) student abilities and effort and (2) the structures that influence students' opportunities.

But, Warren noted, there is considerable debate about how opportunity to learn might be measured. Possibilities for which data are available include class size; pupil-teacher ratio; number of days and hours of instruction per year; teacher quality, perhaps based on value-added measures; or measures of facilities or such offerings as Advanced Placement and International Baccalaureate Programs. These are all proxies rather than direct measures, Warren observed. In his view, further work is needed to determine which measure or combination of measures best captures this essential aspect of schooling, and he finds it "problematic that such a fundamental concept has received so little measurement attention."

ISSUES FOR THE K-12 INDICATORS

Though the panelists were all given the same charge, they approached it in somewhat different ways. Each considered the indicators they chose as a set, intended to cover at least one aspect of the status of education. Some focused only on one of the three aspects defined in the committee's framework; see Appendix C for a list of the indicators arranged by presenter. Thus, individually, each indicator tells a story, Mark Dynarski noted, but taken as a group they should reveal more. For example, if the teacher-student ratio is changing over time, that suggests students may be getting more or less time with teachers. But if that indicator is examined in light of other indicators, of teacher quality or of other factors that may influence the ratio, for example, a more nuanced picture can emerge.

Though the workshop was not intended as a vehicle for making a final selection or recommendations of indicators, discussant Henry Braun encouraged the panelists to use the initial suggestions each had made to consider the characteristics the system as a whole should have. The indicators put forward, he suggested, could easily have been suggested 10 or even 20 years earlier, and he wondered whether further thinking would be needed to ensure that the indicators ultimately chosen will support important contemporary goals for education. The discussion addressed this question from several angles.

College Readiness? Broader Goals?

Several discussants also wondered whether the indicators suggested reflected a sufficiently ambitious vision of what public education can accomplish. One noted that the indicators chosen for a similar project covering European countries (Hoskins, Cartwright, and Schoof, 2010) are based in fundamental social goals: economic security, social cohesion, and sustainability. The United States might have somewhat different goals, the participant added, such as promoting participatory democracy or a balance between individual freedom and responsibility, but the set of indicators chosen could be conceived as measures of how well the system is meeting such goals.

Another participant suggested a different sort of educational purpose that also could provide a conceptual underpinning for the indicators chosen. Beyond the specific skills and knowledge education should impart, he explained, is the idea that education should help children develop images of "possible selves" as they gain understanding of what their options are. Prior efforts to "rethink" high school, he added, such as the Youth Act of 1980[20] or the School-to-Work Opportunities Act of 1994,[21] have not yielded the desired changes. He hopes that current thinking focuses less on the distinctions between college and career preparation and more on helping students prepare to grow and change.

---

[20]For a description, see http://www.eric.ed.gov/PDFS/ED195851.pdf.
[21]For a description, see http://www.fessler.com/SBE/act.htm.

Several presenters pointed out that the skills needed to succeed in the 21st century are not necessarily different from those needed in the past. While adaptation—such as taking advantage of advances in the cognitive and behavioral sciences—is important, it may also be important to recognize that even if ideas have been under discussion for a long time, the system may not yet have succeeded in implementing them. In particular, Pianta noted, possibilities for measuring aspects of education that may have been recognized as important but were not easily captured using large-scale standardized tests have opened up considerably.

It is important to recognize, though, Allensworth noted, that "we have ratcheted up expectations for students" significantly in the last 10 to 20 years. "Almost all schools now say they want their students to leave school college ready," she explained, and this is a significant change from the way things were a generation ago. Tracking student grades is useful, she added, precisely because they capture skills that cannot be measured on tests but are important for both postsecondary schooling and work, such as the ability to get things done, to work in groups, and to solve problems. "There's not a whole lot of evidence out there, she added, "that what you need to be career ready is so different from what you need to be college ready."

Nevertheless, Dynarski observed, "one could read the data and say the K-12 system has just moved too slowly [so it is putting students into] the pipeline who aren't ready for colleges that have adapted more quickly. The empirical evidence is pointing to the shortcomings not of colleges but of the K-12 system."

What Might Be Missing?

Discussion highlighted a few issues that were not raised in the suggested indicators. One participant noted the relative lack of emphasis on contextual factors, particularly in comparison with the suggested preschool indicators. "Presumably the family and the community are just as important for school-age students as for kids from zero to age 5," this person remarked. Another noted that "education is actually coproduced with parents," but that parental inputs were not addressed by any of the indicators. The important role parents play is explicitly addressed in the early childhood context, this person observed, but "over time we assign more responsibility to the education provided by schools." It would be possible, this person suggested, to develop a composite indicator using, for example, parents' reading to children, helping with homework, or helping children prepare for college, to gauge involvement.

It is difficult to address equity issues with national indicators, Mark Dynarski noted, because national statistics may disguise significant variation. For example, steady growth in per pupil spending, on average, may result because affluent school districts are increasing their spending a considerable amount while poorer districts are not. It might be possible to construct an indicator of the variance in the states' spending, perhaps as a share of their own budgets. But even this indicator would not capture other kinds of inequity, such as the distribution of up-to-date buildings, experienced teachers who are teaching the subject in which they were prepared, and other factors. Participants noted that measuring spending at the state level, as well as the national level, would add important information about equity and other issues.

The United States lags behind some other industrialized countries in college participation rates and college attainment, observed William Tierney. In 2006, 40 percent of adults aged 25 to 64 had earned a college degree, putting the nation third in international comparisons (Organization for Economic Co-operation and Development, 2009, 2010). The nation ranks 10th in terms of the percentage of the population that enters college (64 percent in the United States), and 14th in college graduation rate. Indicators for this educational stage could provide better understanding of who attends college, the benefits that college may offer, and college students' experiences and outcomes, but there are challenges.

There are more than 6,600 postsecondary institutions in the United States, of which approximately 4,400 grant degrees.[22] Nearly 1,700 are 2-year institutions, and 2,700 are 4-year institutions. A small but growing segment of the higher education landscape is for-profit institutions, and distance learning opportunities are expanding as well. Within these basic categories, institutions vary enormously in mission, size, population served, resources, and many other characteristics. Making valid and fair comparisons across different types of institutions is a key challenge, and this was a theme throughout the presentations and discussion of possible indicators to monitor the progress of higher education. The indicators suggested by presenters are listed in Table 4-1.

CHARACTERISTICS OF INSTITUTIONS, SERVICE PROVIDERS, AND RESOURCES

Graduation and Retention Rates

The importance of tracking graduation rates was highlighted by most of the presenters, who also offered a variety of comments.

Kevin Dougherty noted that the best way to measure completion has been the subject of considerable debate. The National Center for Education Statistics collects completion data through the Integrated Postsecondary Education System (IPEDS), but it does not currently provide data on part-time students (roughly three-fifths of community college students attend part-time), those who enroll after the fall, or those who transfer (Committee on Measures of Student Success, 2011). "IPEDS is a good dataset but it is

---

[22]For details, see http://nces.ed.gov/fastfacts/display.asp?id=84.

TABLE 4-1 Indicators Suggested for Higher Education

CHARACTERISTICS OF INSTITUTIONS, SERVICE PROVIDERS, AND RESOURCES
<ul><li>Graduation and retention rates, disaggregated to capture community college students and other nontraditional students, and perhaps also financial aid status, family income, and need for remediation at the time of matriculation</li><li>Measure of the highest level of education attained by students 10 years after they first enrolled in a postsecondary institution</li><li>Transfer rates—students who successfully transfer from a community college to a 4-year institution or proportion of students who graduate or transfer within 4 to 6 years of normal completion time</li><li>Educational progress rates, such as: measure of proportion of students college-ready at matriculation; percentage of students who persist through graduation; completion rates for remedial coursework and progression to college-level coursework; cumulative credits earned; or an indicator tracking students K-highest level of schooling in which they enroll</li><li>Preparation for careers and job placement, using employment rates and salaries at 1 and 5 years postgraduation</li><li>Research and development activity, using, e.g., spending on research and development, number of patents secured, or income earned through licenses; indicator for humanities and social sciences also needed</li></ul>
INDIVIDUAL OUTCOMES
<ul><li>Job placement and earnings</li><li>Learning outcomes, such as: cognitive skills or functioning, occupational competence and preparedness; civic awareness and responsibility, global and intercultural competence, moral reasoning</li></ul>
CONTEXT
<ul><li>Navigational capital or understanding of college access and success process</li><li>Participation in different kinds of colleges and programs, including distribution of students by key demographic characteristics (e.g., gender, race/ethnicity, family income, disability status, and age) across different types of postsecondary institutions and higher education outcomes</li><li>Net cost and affordability for families—could include net cost of tuition and fees, minus grants, disaggregated by family income, or average student's loan burden relative to starting salary</li></ul>

important that it be amplified," in Dougherty's view, and he noted that such proposals are currently under consideration.[23]

Lashawn Richburg-Hayes endorsed this view and emphasized the importance of including in the indicator system *disaggregated data that captures community college students and other nontraditional students, and perhaps also financial aid status, family income, and need for remediation at the time of matriculation.* These are important points to track, in her view, because they reflect groups who have the greatest obstacles to success. Laura Perna also noted the challenges of capturing the variation in students and their differing pathways through institutions. She suggested including a basic measure of *graduation rates for full-time students at the institution in which they first enrolled*, but also including tools on the indicators website that allow users to disaggregate the data to reflect graduation rates for different types of colleges and universities and for students with different demographic and academic characteristics.

Another challenge, Dougherty noted, is to decide what time window to use for completion. The current standard is to look at students who graduate within 150 to 200 percent of what is regarded as a normal time for completing the degree. But since so many students attend part time, many take much longer than that. It might be useful, he suggested, to either extend the window or to have several windows, to capture students who stay enrolled or re-enroll. Perna advocated including a *measure of the highest level of education attained by students 10 years after they first enrolled in a postsecondary institution*, as a way of addressing this concern.

Tierney cited both graduation and retention rates (the number of full-time students who return each year) as important, and he also noted another challenge to consider with this indicator. "If graduation is the criterion," he commented, "the for-profits know how to do that—they will graduate students." For that reason it is important to include retention as well, since some students may need more time to meet requirements.

Transfer Rates

A related and equally important measure, for Dougherty, is of *students who successfully transfer from a community college to a 4-year institution without earning a degree at the community college.* However, he added, community college graduation rates usually do not include students who transfer without having first received an associate's degree—a complete picture of the contribution community colleges make would include these data. Such data are available in many state longitudinal data systems and also from the National Student Clearinghouse (an organization that collects data from more than 3,300 participating colleges and universities). Richburg Hayes also called for a measure of the *proportion of students who graduate or transfer within 200 percent (4 years) or 300 (6 years) percent of the normal time to completion.*

---

[23]Dougherty also noted that the College Board is currently developing a website that will make a variety of outcomes indicators for community colleges publicly available.

Educational Progress Rates

Students' progress as they move through higher education was important to several of the panelists, and they suggested several possible indicators. For Lashawn Richburg-Hayes, it is important to begin with a baseline—*a measure of the proportion of first-time, first-year students who are college ready at the time they matriculate*. This is different from the measures of college readiness that might be used to assess the effectiveness of K-12 education, she explained, because many students—particularly those enrolling in community colleges—are not beginning their postsecondary work immediately after high school.

More than half of community college students matriculate 5 or more years after high school, and their average age is 26 to 28. Even if those students were college ready at the time they completed high school, she observed, they are likely to have forgotten "the trigonometry, the algebra, even the fractions, for that matter, because they did not use them" after high school. In order to fairly assess community colleges, then, she believes it is important to understand "what it is they are starting with."

Similarly, Laura Perna included a measure of college readiness to monitor the pipeline of students entering college, such as the *percentage of 9th graders who graduate from high school, enter college, persist through the first year of college, and ultimately graduate*. This should be a K-12 indicator, she noted, but since she did not see consensus on this point in that discussion, she included it as a higher education indicator. She noted that college entrance examinations are not ideal measures of college readiness.

Looking next at what takes place during postsecondary schooling, Dougherty noted that several indicators can show whether or not students are on track to complete a degree. These can also be useful to policy makers because they can help identify the "points of blockage students are running into," he added. One is a measure of *completion rates for remedial or developmental programs and whether students go on to complete a college-level course in the area in which they received remediation*. Twenty-five states collect these data, and many also collect data on the pace at which students accrue credits toward a degree, and the time it takes them to earn a credential (García and L'Orange, 2010). Richburg-Hayes also advocated measuring completion rates for developmental education requirements.

William Tierney also cited the importance of measuring institutions' capacity to meet students' needs for remediation. While needs vary across types of institutions, he noted, "We need criteria that work across the postsecondary sector" to measure this because students at every type of institution come in needing remediation.

Richburg-Hayes also cited the importance of measuring institutions' effectiveness at remediation, and suggested using "*the proportion of students who were not prepared for college-level work upon matriculation who pass the developmental course requirements within three semesters*." Until "you deal with the missing skills," she noted, "it is not reasonable to think about graduation rates and transfer rates."

Richburg-Hayes also advocated another indicator of students' progress through postsecondary education, the *cumulative total degree-applicable credits students earn*. Community college students tend to approach college differently from the way more traditional students do, she explained. They may have many different reasons for attending and usually integrate their coursework into work and family commitments differently than do 4-year students. When students need considerable remediation, they may need to delay the courses required for the degree. These students take longer to meet graduation requirements and are the most likely to drop out before earning a degree.

"Since remediation barriers have been identified as a key deterrent to graduation," she concluded, tracking the rate at which students are meeting actual degree requirements and other goals is important. Doing so is complicated, however, by the fact that institutions have varying methods for assessing the need for remediation as well as varying courses and procedures for providing it.

Perna looked at progression in a different way, suggesting an indicator to *track students from pre-K through the highest level of schooling in which they enroll*. These data could potentially be linked to occupational data, she noted, which would provide important insights about the short- and long-term benefits of different types of preparation and postsecondary pathways. She noted that state longitudinal data systems and the National Student Clearinghouse have begun to track students more comprehensively, so the pre-K to postsecondary indicator may not be an impossible dream. She agreed that measuring college learning is important but believes that the filed does not currently have good ways to measure it.

Preparation for Careers and Job Placement

Regardless of an institution's mission and the characteristics of the population it serves, it has a responsibility to provide students with the skills they will need to meet their goals. It is important to monitor how well they are doing so, "whether for cosmetology or engineering," William Tierney noted, and whether schools are placing students in jobs comparable to their education and training.

Perna addressed this in a more concrete way, suggesting a measure of *the economic benefits to individuals and society provided by institutions, in terms of employment rates and salaries in the short and long term (e.g., at 1 and 5 years postgraduation)*. She also cautioned that use of an economic indicator of benefits reflects the reliance on data that are available; numerous other benefits also result from higher education but are less easily quantified and measured.

Research and Development Activity

Higher education institutions have responsibilities beyond educating students, observed Dougherty, which should also be monitored. Obvious measures, such as *spending on research and development, number of patents secured, or income earned through licenses*, are important but capture primarily activity in the natural and biomedical sciences. Indicators for *the humanities and social sciences* are needed as well, in his view.

INDIVIDUAL OUTCOMES

Job Placement and Earnings

For Dougherty, preparation for careers should be considered as an outcome for both individuals and institutions. Federal and state employment data are important sources of information on outcomes for students once they leave higher education, Dougherty explained. There are difficulties with interpreting these data, however. Labor markets are very volatile, he noted, and conditions may vary markedly from one state or region to the next. Thus, employment rates and earnings may be more related to labor market conditions than to the effectiveness of higher education. In his view, an indicator of *job placements and earnings* will be important but the data must be interpreted carefully.

Learning Outcomes

"The real hole in what we know about higher education is that, for all intents and purposes, we are unable to say anything at the state or national level about *what* students are learning and *how much* of it they are learning," argued Patrick Terenzini, and that was the focus of all of his suggested indicators. A widely cited regular report on higher education produced by the National Center for Public Policy and Higher Education, *Measuring Up,*[24] he noted, provided measures of college preparation, participation, access, affordability, and completion, for example, but reported little about what students learn at postsecondary institutions. There has been some interest in expanding the sampling for the National Assessment of Adult Literacy to provide state level data for a few states, he added, but in his view, the literacy skills that are assessed are "quite basic." At some point we need to look for measures of higher levels of development," he noted.

The idea of expanding NAEP to cover postsecondary education has been proposed, Terenzini noted, but he acknowledged the difficulty of measuring so diffuse a construct as college learning. There is very little consensus on what students should learn "beyond some fairly high-flying, abstract statements about things like critical thinking," he added, which is widely supported "until you start trying to define it." Nevertheless, he noted that there is a considerable body of research on higher order thinking, occupational competence, and civic awareness and participation that provides the basis for possible indicators of college learning (Pascarella and Terenzini, 1991, 2005). Terenzini suggested several areas that would be worth probing, though there are obvious indicators for only some of them. Acknowledging the difficulty of measurement in this area, he noted that "we have to start somewhere." He suggested several indicators:

- *cognitive skills or functioning among college students and adults.* This would include critical thinking, problem solving, synthesizing, the ability to evaluate evidence, and the ability to exercise judgment, and could be evaluated using such available measures as the ACT's Collegiate Assessment of Academic Proficiency

---

[24]For more information on these reports, see http://measuringup2008.highereducation.org/; however, these reports were discontinued after 2008.

or the Collegiate Learning Assessment.[25] In Terenzini's view, these would be the best available options, but he acknowledged that using either as an indicator of college learning would be challenging.

- *occupational competence and preparedness for advanced practice in specific fields*. For this, Terenzini would use data from licensure examinations (where available) in such professional fields as nursing, teaching, physical therapy, and engineering, as well as measures of preparedness for graduate study, including the Graduate Record Examination, the Medical College Admissions Test, and the Law School Admissions Test. American College Testing's WorkKeys program, which evaluates 2-year college students' preparedness in such areas as applied mathematics, locating information, and reading for information, could also be used.[26]

- *civic awareness and responsibility*. Cultivating a sense of membership in a community and the will to participate in that community has been a valued goal in higher education since its beginnings in the United States, Terenzini noted, and there are currently at least two instruments for it. The National Conference on Citizenship is developing a Civic Health Index, and the annual Bureau of the Census Current Population Survey supplements recently began including questions about volunteering, voting, and other indicators of civic health.

- *global and intercultural competence*. This indicator would include, for example, the ability to understand people who have different cultural backgrounds, both within the United States and worldwide, and the ability to work in groups.

- *moral reasoning*. Terenzini believes that it is an important goal of higher education to cultivate not any one set of moral stances, but rather the capacity to reach judgments of one's own about right and wrong—as opposed to relying exclusively or primarily on religious tradition, parental authority, or other authorities for such judgments.

CONTEXT

Navigational Capital, or Understanding of College Access

Along with college readiness (whether as a K-12 or higher education indicator), Dougherty argued, it is important also to have an indicator of *navigational capital*, which he defined as students' knowledge of the college access and college success process (see Yosso and Solorzano, 2005). A few researchers have explored what happens to students when they get to community college or a for-profit school, he noted (e.g. Rosenbaum, Deil-Amern, and Person, 2006). The admissions and financial aid process; college academic requirements, organizational procedures, and expectations; and curricular pathways may all be unfamiliar and daunting for many students.

---

[25]For more information about these programs, see http://www.act.org/caap/ and http://www.collegiatelearningassessment.org/.
[26]A participant noted that the National Adult Literacy Study includes a measure of technological literacy that could be used to characterize skills in that area by age cohort.

Although he knows of no good way to measure how well students understand what is involved in the college process, he believes it is an important measure to consider as an aspiration. "Providing opportunities is not enough," he commented, and the knowledge needed to successfully navigate this very complicated process "is very socially stratified." That is, this factor may be especially important for community college students. Since many of them are the first in their families to attend college, they are less likely to have family members with knowledge about the higher education system who can advise them.

Participation in Different Kinds of Colleges and Programs

The structure of the higher education system, in terms of *the array of 2- and 4-year institutions, as well as offerings in different fields,* should also be tracked, in Dougherty's view, because there is evidence that outcomes for students vary according to the type of institution they attend and the programs they complete (Long and Kurlaender, 2009; Pascarella and Terenzini, 2005). For example, he noted, it would be useful to show breakdowns of college choice and major by students' family background. Such information is likely to illuminate findings about students' later outcomes. Perna also addressed this issue, framing it as a measure of equity, in terms of the *distribution of students by key demographic characteristics (e.g., gender, race/ethnicity, family income, disability status, and age) across different types of postsecondary institutions and higher education outcomes.*

Affordability

A related access issue is *net cost and affordability for families,* Dougherty added, which he suggested could be measured as the net costs of higher education in relation to average family income (National Center for Public Policy and Higher Education, 2008). Tierney also cited this issue, calling for a measure of debt incurred in relation to the income graduates can earn, which might also be treated as an institutional indicator. Perna addressed this issue as well, advocating a measure of the *net cost of tuition and fees, minus grants, disaggregated by family income.* Like Tierney, she believes it is also important to have a measure of the magnitude and manageability of student debt, perhaps the *average graduate's loan burden relative to starting salary.*

ISSUES FOR HIGHER EDUCATION INDICATORS

To open discussion of the many ideas put forward by the panelists, Lisa Lynch identified four basic categories she saw in the suggested indicators: college readiness, affordability, access, and some way of measuring the value that higher education offers to individuals and society. She also noted that relatively little was said about technology—though she believes it is transforming the delivery of higher education—or about the changing demographics of the college-going population. Options for online learning, for example, may mean that students, faculty, and the institution itself could be located in entirely different places, which may complicate the collection of data and the possibilities for making comparisons across institution, student population, geographic location, and

40

time. At the same time, she added, the traditional student, who enters college at age 18 and attends more or less full-time, may soon be in the minority. Thus, she noted, data collection will need to adapt to the growing proportion of students who do not fit this mold.

In the discussion, participants expanded on these and other issues.

A Complex Sector to Measure

Richburg-Hayes noted that many of the indicators suggested for the higher education stage are not currently available and also that many will be difficult to standardize. Efforts are under way to begin collecting much of the necessary data, but the lack of standardization is a more difficult problem to solve, in her view. In many cases, colleges have different definitions for basic terms that will make it difficult to compare data. For example, for one college, "degree-seeking students" means all who enroll, while for another, that group may include only students who have formally submitted documentation of their intention to pursue a degree offered at the community college or transfer to a 4-year institution. A comparison of the graduation rates of these schools would need to factor in this difference. The purpose and rigor of coursework also vary considerably.

There are few metrics that are common across colleges and can be used to make valid comparisons across institutions and over time, she added. The postsecondary sector lacks a counterpart to the National Assessment of Educational Progress, which provides a standardized measure of proficiency in academic subjects that permits comparison of student cohorts across time and among of states. "We would not necessarily compare Harvard's graduation rate with those of local community colleges," she added, "but that is essentially what we are talking about doing with indicators." Similarly, even though Harvard might accept a transfer credit for biology 101 from a community college, the course is likely to be qualitatively different from Harvard's version.

This is not just a methodological issue, she continued: "There is an underlying conceptual problem with having indicators for higher education that we would need to tackle to avoid the apples to oranges problem." This is not an insurmountable problem, in her view, but one that cannot be ignored. Others agreed. One commented, "I am not even sure that any two or three institutions can really be put in the same stratum." It might be more useful, this participant suggested, to compare institutions by outcomes for particular demographic groups that are more comparable. "Being able to disaggregate to an appropriate level is essential to the validity and utility of the data," this person added. Another noted that, without a new approach, "We will continue to praise the successes of the advantaged, elite institutions, at the expense of all the others."

Focus on For-Profit Institutions

Many people question the value of for-profit institutions, noted Tierney, but in his view they have an important role. "If we want a small or relatively elite system," he commented, then they may not have a role, but "if we want to expand enrollment, for-profit [institutions] have to play a role." He believes that for-profit institutions are not going to take over higher education, but he added that few people would likely have predicted a decade ago that 12 percent of college students would be attending for-profit institutions (this figure includes institutions that do not grant degrees or certificates). He sees a steady increase in the number of people pursuing higher education, as the demand for a better educated workforce grows nationwide. If participation rates in higher education are to increase, institutions will need to focus on two populations that have traditionally had low rates of college attendance: low-income, first-generation students of color and working adults. Public postsecondary institutions do not currently have the capacity to expand enough to serve these groups adequately. For instance, he predicts that student participation in higher education in California is likely to increase by 100,000 per year every year for the next decade, while tight state budgets mean that funding for public institutions is either flat or declining.

Public postsecondary institutions will not be able to meet the need, and, in his view, for-profit ones will be needed to fill the gap. In 1967, fewer than 22,000 students, or less than one-third of 1 percent of postsecondary enrollment was in for-profit, degree-granting institutions, while today that number is 1.2 million, or 6.5 percent of the total (see Hentschke, Lechuga, and Tierney, 2010, and Horn, 2011, for data and other information on for-profit institutions). The for-profit sector is the fastest growing sector of higher education, and these institutions offer more online courses than their public and private counterparts do. They also serve high percentages of first-generation college students, students of color, and working adults, Tierney added.

For-profit institutions vary significantly in quality, however. This sector is growing very fast and, as in any industry that is growing fast, "there are some fly-by-night operations," Tierney added. Even though the for-profit institutions may resist the push to measure their outcomes, they will ultimately benefit, in his view, if the distinctions between legitimate institutions and low-quality ones becomes more evident. "The danger if we don't implement this correctly is that those students who need it the most—students of color, first-generation students, and working adults—will be left out," he concluded.

Purposes for the Data

The diversity of the higher education sector is just one reason that many presenters and participants emphasized the importance of being able to disaggregate the data collected for many of the suggested indicators. There are many possible audiences for these data, Perna observed—the general public, consumers, government, employers, and institutions themselves, for example—and each may have different uses for them. Institutions might want to compare themselves to their peers as they pursue improvement, while other users might be more interested in whether funds expended on higher education are being used effectively and efficiently. Each purpose might point to different sorts of indicators, she added. In her view, the overarching goal should be monitoring the educational attainment of students at all levels.

Tools for disaggregating the data available on the national indicators website could, however, allow for multiple uses. Many participants suggested that it should be possible to explore data by type of institution (using more precise categories than, for example, 4-year and 2-year) and also by type of student (e.g., full time, part time, and by such demographic characteristics as race, gender, and socioeconomic status). The possibility of isolating students who do and do not receive Pell grants, and those who do and do not enroll in developmental courses, for example, would provide a way of exploring socioeconomic factors, noted one participant. Another issue to consider is what would be necessary to make valid temporal and geographical comparisons, a participant observed. In order to support international or even state-by-state comparisons, it is important to control for differences in the composition of students populations across the entities compared. Otherwise, this participant noted, "you could generate data that carry a lot of weight in the media but don't fairly represent what is going on.

Equity Issues

Equity issues were not as prominent in the indicators suggested as some participants expected, and they raised several points. The legal system has played a large role in determining approaches to equity in the K-12 sector, one noted, but that has not been the case in the postsecondary setting. For higher education, he noted, "whether public or private, it is a moral obligation." Affirmative action has been significantly scaled back, this person observed, so new ways are needed to consider the equity issues reflected in differential rates of participation in higher education, access and availability, and retention and graduation. There are two distinct issues, another added. One is the idea that there is a responsibility to address past inequities to minority and other groups, and ensure that those inequities do not persist, by means of particular attention to the adequacy of public education for all. The other issue is the right of educational institutions to take particular steps to pursue diversity for the sake of its educational value.

Both issues could be illuminated by data on the distribution of different types of students along different pathways, on transfers, course-taking, and the labor market value of different pathways, and one participant noted that such data are incomplete at present. It appears that there is a disproportionate concentration of students from low-income families in for-profit institutions, one person noted, wondering if data could be collected to shed light on questions such as, "Who are these kids? What fields are they going into? How do they fare?"

Another agreed, noting that simply documenting the differences in employment, experiences, earnings, and other benefits that may accrue from higher education, across groups and pathways, is important. "Part of what a national indicators system should do is lay out what is happening," this person observed. "Then others can ask why, which characteristics and forces are contributing to differences across groups."

Institutional Quality

Some of the suggested indicators address a related issue—what students actually learn in college—but none directly address the quality of institutions, one participant noted. There are possible measures, this person added, such as exposure to curriculum relevant to particular goals, instructional quality, student engagement, diversity, institutional culture, for example. It is relatively easy to track more concrete indicators, such as financial resources, accomplishments reflected in prizes awarded to faculty, or licensure awards for faculty inventions and those are important, several people noted. But in the K-12 sector, a participant commented, it has become clear that it is not enough to look at how much money a district has: one has to see how it is flowing into individual schools and even classrooms. Similarly, he went on, in the postsecondary sector, "we need to look at the capacity of institutions for organizational learning, because so much of what we are talking about is organizational change."

As the sector grows and grows more diverse, the challenge of measuring its status, its quality, and what it offers to those who consume it, will become more complex, participants suggested. "We talk about it as if it were a single, coordinated system," one noted, "but it doesn't operate as a system—it is a nonsystem."

5
Indicators for Adult Postsecondary Education and Training

"Adult postsecondary education and training" is a somewhat ambiguous and elastic category, noted Marshall Smith. It might encompass, for example, coursework at a community college that does not result in a certificate, studying for an occupational license, or informal online learning an individual does independently to boost his or her resume. The workshop focused on two broad categories: continuing education opportunities that allow adults to pursue areas of personal interest or build skills and knowledge for their careers, and formal adult basic education programs for adults who have not earned a high school diploma. The indicators suggested are listed in Table 5-1.

Allan Collins emphasized the importance of continuing education opportunities, noting that "we are in the midst of a major transformation in education in the United States," that is mostly happening outside of school. Workplace simulations, business education in virtual high schools and colleges, for-profit learning centers, a proliferation of online courses and technical certifications, older workers starting second careers—all of these are part of a new education frontier for adults that is largely made possible by technology.

There is a growing demand for continuing education, noted David Stern. In a typical U.S. lifespan, the number of years spent in paid employment is approximately three times the number spent in formal schooling, so education that takes place in the context of employment meets a significant need. Stern added that such opportunities provide a clear benefit to workers (in terms of higher wages, for example), though the benefit to employers is less clear because employees may leave jobs before their employers have recouped the cost of training they provided. Thus, employers may be providing fewer such opportunities than would be optimal, Stern noted. Moreover, research has shown that white and highly educated employees receive more such opportunities than their African American, Hispanic, and less well-educated peers.

Employment-based education can best be seen as a continuum, Stern added, ranging from formal on-the-job training for specific purposes to spontaneous, unscheduled learning that takes place during informal interactions or online. For example, many employers have developed online learning systems that help employees acquire necessary skills or knowledge on an as-needed basis. Between the most and least formal sorts of work-based education, Stern added, is a very interesting array of semi-structured activities and practices—such as job rotation (where individuals experience different responsibilities in a workplace); skill-based pay; and cross-training (teaching an employee the skills needed for a job other than the one for which he or she was hired)— that are not classroom-based. In this context, basic indicators of availability and

TABLE 5-1 Indicators Suggested for Adult Postsecondary Education and Training

CHARACTERISTICS OF INSTITUTIONS, SERVICE PROVIDERS, AND RESOURCES
• Types of on-the-job training provided by employers, using a survey on such questions as whether employers support individual and group innovation and whether employees have individual or collective learning plans • Percentage of adults aged 25 or older who enroll in postsecondary education or training and who earn credentials: • Percentage of adults 18 or older who have left high school without a diploma but who obtain a high school diploma, GED certificate, or equivalent • Percentage of low-skilled adults who obtain a postsecondary or occupational certificate, credential, or degree • Percentage of instructors in adult education programs who are certified in an adult education field
INDIVIDUAL OUTCOMES
• Percentage of adults aged 18 or older who pass the examination to become naturalized citizens or • Percentage of adults aged 18 or older who vote for the first time • Participation by employed individuals in on-the-job training • Percentage of the adult population who believe they know how to learn and are motivated to exercise this skill
CONTEXT
• Adult literacy rates • Spending on adult education as a percentage of gross domestic product (GDP) • Shortages of key skills in the labor market • Health of the support network for adult learning associated with occupational interests

participation make sense, but it is likely too difficult to develop a standardized set of outcome measures for content that is "so varied, ephemeral, and often so specific to a particular learning environment," Stern observed. Moreover, it is important for adults to be able transfer knowledge from one realm to another, but this capacity is difficult to measure.

Formal adult education, in contrast, is largely funded by the U.S. Department of Education (under Title II of the Workforce Investment Act of 1998). This system is focused on assisting adults who have not earned a high school diploma or whose reading, writing, or mathematics skills are below the secondary level. Most adult education services are provided through local education agencies, community colleges, libraries, community organizations, and correctional facilities, and also include programs to engage adults in the community, support families, and promote work and learning (U.S. Department of Education, Division of Adult Education and Literacy, 2011). Other federal programs (including Title I of the same law and a job training program administered through the U.S. Department of Labor), as well as those of foundations; state and local governments; and private, for-profit organizations, also provide programs that serve low-skilled adults.

These services, Judith Alamprese explained, span multiple ages and stages of learning, and it is challenging just to count how many people participate, estimate the cost, and track participants' outcomes. The Department of Education's program serves approximately 2.5 million people annually, she noted, with the goal of preparing adults for postsecondary education, training, and employment (U.S. Department of Education and Division of Adult Education and Literacy, 2010). The need is great. Roughly 30 percent of adults aged 25 to 44 (or 30 million people), she noted, do not have a high school diploma (U.S. Census Bureau, 2010), although economists forecast that 63 percent of future jobs will require more than a high school diploma (Carnevale, Smith, and Strohl, 2010).

It is important that a new indicator system capture all the facets of this sector, Collins noted, but it is fast changing and thus intrinsically difficult to monitor.

CHARACTERISTICS OF INSTITUTIONS, SERVICE PROVIDERS, AND RESOURCES

Types of On-the-Job Training Provided by Employers

A basic measure of how much on-the-job training is available is important, in Stern's view, and could be obtained relatively easily through a supplement to the ongoing Current Employment Statistics Program run by the Bureau of Labor Statistics. That survey of establishments is conducted monthly, and a supplement could be included yearly to ask about whether employers offer classroom instruction, online resources, or other educational opportunities, and about the proportion of employees involved, the duration of the educational opportunity, and how frequently it is available.

Marshall Smith also advocated a measure of the learning opportunities provided by employers, suggesting that surveys or other data collection methods be used to ask *whether employers support individual and group innovation* and *whether employees have individual or collective learning plans.* He noted that case studies of successful organizations indicate that they are places where employees continually learn and improve their work. Increased attention to this aspect of the work environment, he suggested, could encourage all employers to emulate the healthiest ones.

Participation in and Credentials from Postsecondary Institutions

Judith Alamprese suggested three indicators of the availability of adult basic education options because there is a range of circumstances that she believes should be captured. She noted that there are "denominator problems" with questions about the rates at which adults who need adult education services enroll in and complete such programs—that is, decisions about whom to count. Nevertheless, one possible indicator would be the *percentage of adults aged 25 or older who enroll in postsecondary education or training.* The second would be the *percentage of adults 18 or older who have left high school without a diploma but who obtain a high school diploma, a GED certificate, or the equivalent.* The third would be the *percentage of low-skilled adults who obtain a postsecondary or occupational certificate, credential, or degree.*

Sources for such data include state longitudinal databases that track students from adult education through postsecondary completion, the National Center for Higher Education Management Systems, the U.S. Citizenship and Immigration Service, and the U.S. Department of Education.

Certification of Program Instructors

Few states have a required certification in the field of adult education, Alamprese noted, and few data are currently available to document the quality of instruction that is delivered in adult basic education programs (Smith and Gomez, 2011). However, many states are beginning to develop certification systems, and the U.S. Department of Education has planned to begin collecting data on the quality of teaching in adult education, so there will soon be baseline and progress data (U.S. Department of Education, Division of Adult Education and Literacy, 2012). Thus, one possible indicator would be *the percentage of instructors in adult education programs who are certified in an adult education field.*

INDIVIDUAL OUTCOMES

Becoming a Naturalized Citizen and Voting

Civic and social outcomes are key goals for formal adult education programs, Alamprese noted, and she suggested two possible indicators: (1) *the percentage of adults aged 18 or older who pass the examination to become naturalized citizens* and (2) *the percentage of adults aged 18 or older who vote for the first time.* She noted that other indicators could also be used, but that data for her proposed indicators are available from the U.S. Citizenship and Immigration Service, the U.S. Department of Education, and other sources.

Participating in On-the-Job Training

In addition to assessing education opportunities employers say they offer, in Stern's view, it is also important to track *the on-the-job training opportunities employees say they have experienced.* This could also be done by adding questions to an existing survey, he noted, in this case, the Current Population Survey conducted jointly by the U.S. Census Bureau and the Bureau of Labor Statistics.

Knowing How to Learn

It is important to understand how prepared the adult population is to take advantage of learning opportunities, Smith argued, because being prepared in this way is essential to success in a modern, high-skill job. He proposed an indicator of *the percentage of the adult population who believe they know how to learn and are motivated to exercise this skill.* He suggested that a survey question such as "how would you attempt to solve an unfamiliar problem?" or a short essay question could be used to measure this attribute, and that attention to this factor would encourage schools to focus on it.

CONTEXT

Adult Literacy Rates

A critical basic question, for Alamprese, is how literate adults in the United States are, so an indicator of *adult literacy rates* is needed. The National Assessment of Adult Literacy (NAAL) has measured English literacy among U.S. adults 16 and older, as well as background characteristics, since 1992. The most recent data, from 2003, show what 43 percent of adults scored at the below basic (14 percent) level or basic (29 percent) levels (Kutner et al., 2007). Another assessment currently under way, the International Assessment of Adult Competencies, will provide international comparisons of adult literacy rates (Lemke and Gonzales, 2006).

Spending on Adult Education

A primary question, in Alamprese's view, is whether sufficient resources are being expended to support high-quality services for adult education. The estimate for 2010, she noted, is that less than 1 percent of the nation's gross domestic product (GDP) was spent on federal, state, local, and private adult education. She said an indicator of *spending on adult education* is needed.

Shortages of Key Skills in the Labor Market

Stern proposed an indicator of *shortages of key skills in the labor market.* While this measure may be more of a labor market indicator, he noted, it is important to understand the demand for employment-related education and training, particularly because the demand evolves with technological and market changes. A measure of job openings in new and emerging occupations, could be used for this purpose, and ongoing surveys conducted by O*NET and online resource for a variety of occupational data, could provide this sort of information.

Support Network for Adult Learning

The organizations and services (e.g., internet resources, local tutors, friends, work colleagues, welfare-to-work training) that can support individuals in pursuing on their own learning that will be of benefit for current and future employment may have an important impact that is not well understood, Smith argued. Surveys could be used to assess *the support network associated with occupational interests that are available in different regions and among different population groups.* He also suggested a related question, regarding people's proclivity to take advantage of opportunities they have. Measuring, for example, how frequently people report having gained knowledge or skills related to their work outside the workplace or using professional networks could be an additional signal of equity issues or cultural differences that need to be addressed.

ISSUES FOR ADULT EDUCATION AND TRAINING

Lisa Lynch opened the general discussion with an overview of key messages from the presentations on adult learning. First, she noted, it is important to erase the sense of a boundary between education and training for adults and to treat both as important learning. Another key issue, she suggested, is that there is so much that we don't know. Literacy assessments, labor force projections, and data on the proportion of national wealth spent on adult education are already available and provide a general indication of the need. These are not sufficient, in her view, to answer important questions about the extent of the need for more learning after formal education is completed, who has opportunities for learning, where those opportunities are housed, and they sorts of deficiencies for which learning opportunities are needed. Nor do they answer questions about the content of opportunities for adult learning, though this is likely to be an especially challenging area to measure. In addition, there are not readily available data on private-sector spending.

Other important questions concern how adult learning is paid for and what sort of return the investment offers to employers or markets. "We don't know enough about the leaning enterprise to be able to say," she added, whether market failure accounts for the apparent shortage of opportunities for postschool learning, or whether it is too expensive, or whether it is too time-consuming.

The brain is much more responsive to learning later in life than was once thought, Lynch added, and advances in cognitive science have expanded appreciation of possibilities for lifelong learning. Yet the thinking about developmental learning that has strongly influenced K-12 education has not had comparable influence on adult education. There are no clear ways of measuring the competence of training and learning institutions that serve older people or whether the individuals who provide the training "actually understand the most effective methods for training and educating adults," she added.

The discussion followed Lynch's lead in focusing on the big picture and centered on two issues: the need for information about low-skilled and undereducated workers and the challenge of data collection.

Information About Low-Skilled and Undereducated Workers

For low-skilled and undereducated workers it takes a lot to come back into the system, noted one participant. A major goal of adult education is to inspire people to continue their own education, but many people have specific challenges that impede their progress. This is a very heterogeneous group, another noted, that includes recent immigrants who may have low English skills and may also lack knowledge of the education and work landscape in the United States. It may also include individuals with learning disabilities, recognized or not, who have not been successful in formal schooling. It would be useful, then, to have indicators particularly focus on these groups and also apply across age groups and life stages.

One participant noted that society has a compelling interest in whether individuals perceive that they have careers, and whether they are progressing along some sort of track. Defining what constitutes a "career," he continued, may be challenging, but sequence is the key word.

Data Collection

While there was little disagreement about the importance of any of the indicators suggested, one participant was struck by the daunting challenge of collecting data and developing indicators. The United States has a number of valuable data sources already, another noted, but different agencies ask the questions, and each one has to do it a little differently, so you cannot actually compare a lot of the data over time. Others lamented that the United States is often missing from international comparisons related to adult learning because data are not available that will work for these purposes. Nevertheless, several stressed that the United States has underused resources, such as the American Time Use Survey that was included in the Current Population Survey. That survey includes such information as participation in educational activities; work-related activities; and organizational, civic, and religious activities.

Participants noted the simplicity of the European Lifetime Learning Indicators, a composite index covering European countries, which covers: learning to know (the formal education system); learning to do (vocational training); learning to live together (learning for social cohesion); and learning to be (learning as personal growth). More important, Chris Hoenig concluded, is that "there is a huge movement worldwide to share best practices on how to define progress in an amazing variety of areas." There is much to be learned from international examples, several agreed.

6
Indicators for Lifelong, Informal Learning

Learning continues across the life span. It takes many forms and occurs in many venues, both within and outside the formal structures of education. Although there is a seemingly infinite variety of possible sources and stimuli for learning, society has an interest in understanding what and how people are learning informally, and the ways such learning can be fostered. By definition, informal learning is not primarily guided by institutions, although schools, museums, and other entities are important sources of opportunities for it. Still, such learning builds skills—from language and cultural norms to technical expertise—that are needed in the workplace. Informal learning is important to the structures of civil society and democratic governance, and it is an important part of fulfilling life.

Both the Canadian Council on Learning and the OECD have recently focused on lifelong learning and have offered approaches to defining and measuring informal learning; the committee considered their models in thinking about education indicators (Canadian Council on Learning, 2010; Werquin, 2010). The Canadian Composite Learning Index and the European Lifelong Learning Indicators both capture different sorts of learning that occur throughout life in school, in the community, at work, and at home. Both are composite indicators, based on domains defined by the UNESCO: learning to know, learning to do, learning to live together, and learning to be. These indicators are designed to support international comparisons for all basic types of learning.

In planning this portion of the workshop, the steering committee recognized that lifelong, informal learning does not fit neatly within the framework used to structure the workshop (see Table 1-1). The range of "institutions and service providers" through which one learns informally is much broader than for the other stages because individuals can choose to engage with a wide variety of both formal and informal resources. Likewise, the "contextual factors" that may influence informal learning are broader and more diffuse than those that directly affect formal schooling. Finally, measures of behaviors, engagement, and outcomes associated with informal learning are difficult to define and obtain. Despite these limitations, the committee judged it critical to address this aspect of education during the workshop, particularly to provide a foundation for future work in identifying indicators.

The committee identified three important topics related to lifelong, informal learning to define the scope of the discussion: political and civic engagement, media use, and cultural engagement. They asked experts in each of these areas to address the same six questions given to the other panelists, while acknowledging the challenges of this request. This chapter is structured around the discussion of the three topics. All the indicators suggested are shown in Table 6-1.

TABLE 6-1   Indicators Suggested for Lifelong, Informal Learning

POLITICAL AND CIVIC ENGAGEMENT
<ul><li>Practices such as voting, boycotting, supporting a political party or candidate</li><li>Quantity and equality of engagement in such practices as volunteering, attending public meetings, working to address community problems, and making charitable donations</li><li>Quantity and quality of political and civic news consumption</li><li>Quality and quantity of engagement with diverse views on civic and political issues</li><li>Opportunities to learn content related to civic and political life</li></ul>
MEDIA USE
<ul><li>Access to media, beginning with measures of household spending on or ownership of media devices and sources</li><li>Media exposure that captures the nature of the content being consumed</li></ul>
CULTURAL ACTIVITIES
<ul><li>Engagement in learning opportunities outside of formal schooling, perhaps using surveys administered after learning experiences</li></ul>

POLITICAL AND CIVIC ENGAGEMENT

Joseph Kahne focused on indicators of political and civic engagement. For him, these two distinct but overlapping types of engagement are among the most important outcomes of education, and he expressed concern that they are in decline. Quoting former president of the University of Chicago, Robert Maynard Hutchins, he cautioned that "the death of democracy is not likely to be an assassination by ambush, it will be the slow extinction from apathy, indifference and undernourishment." To demonstrate this risk, he displayed data collected through a supplement to the U.S. Census showing disparities in the participation of U.S. adults aged 25 and older[27]: in 2008, 74 percent of adults with a college degree voted, 53 percent of those with a high school degree voted, and 31 percent of those with neither degree did so. Rates of voluntarism show the same pattern: 42 percent of college graduates volunteered that year, while just 18 percent of those with a high school degree and 9 percent of those with neither degree did so.

Kahne suggested including an indicator of *political engagement statistics on such practices as voting, communicating with public officials, boycotting, attending meetings where political issues are discussed, and showing support for a political party or candidate.* The indicator might distinguish among being "highly engaged," "engaged," and "disengaged," or indicators could be created for particular acts, such as voting.

---

[27]For more details, see http://www.census.gov/cps/about/supplemental.html.

Another indicator of civic engagement would include *statistics on the quantity and equality of engagement in such practices as volunteering, attending public meetings where there are discussions of community affairs, working to address community problems, and making charitable donations.*

The U.S. Census Bureau already collects valuable information about many of these actions and behaviors, he noted, but current data do not capture engagement that occurs in the digital sphere, such as communicating electronically with elected officials or joining online groups. Asking about whether respondents have written a letter to the editor, for example, is no longer the best way to track engagement, he explained, and tracking media consumption now needs to include new venues. Following the evolving nature of involvement, however, means giving up on some of the longitudinal data. "You have to think about how to handle that—but it doesn't mean you can ignore it," Kahne said.

New media are particularly important influences on the ways people obtain information. Kahne noted that research has shown a close association between reading, watching, discussing, and otherwise engaging with news that concerns societal issues and being well informed about civic and political life (e.g., Deli Carpini and Keeter, 1996; Popkin and Dimock, 1999). But this area also has changed significantly. For example, many people now get their news from friends or members of online groups who share it with them, rather than through self-selected consumption of material published by a particular source. He suggested an indicator of *learning through engagement with civic and political media sources, which would measure the quantity and quality of political and civic news consumption.*

Also important to Kahne is an indicator of *learning through engagement with diverse views on civic and political issues, which would include statistics on the quantity and quality of such engagement.* Kahne noted that exposure to a diverse range of views related to civic and political issues is important. Research has suggested that such exposure can foster an individual's ability to see other perspectives, appreciation for the legitimacy of the rationales put forth by those who disagree, and political tolerance for those with differing perspectives. At the same time, there is some evidence that such exposure may lead to reduced levels of participation (Mutz, 2006), a finding that Kahne suggested needs further investigation.

More general learning was the focus of his final indicator, which concerns such *civic learning opportunities as learning content (e.g., history, economics, sociology, political science) that relates to civic and political life; discussing current events; having open and respectful dialogue about current events; engaging in community service; engaging in extracurricular activities; and participating in simulations of civic and political processes.* Some of these opportunities occur in the context of formal education but others are less formal. Research has suggested that these opportunities promote civic and political engagement, but also that there is considerable inequity in access to such opportunities (e.g., Hart et al., 2007; Kahne, Feezell and Lee, 2012; McDevitt and Kiousis, 2004). However, Kahne noted that such opportunities and their outcomes are not generally measured.

Including these issues in the education indicator list, Kahne suggested, may reinforce the value of civic and political engagement and spur educators and others to place more emphasis on developing it. However, he cautioned that he could not see a ready way to capture the quality of all of these types of engagement, noting that a measure of engagement with news media would be more useful if it captured the quality of the information the media provided. He also stressed the importance of documenting inequities in the participation of different groups, and whether or not groups receive equal support for engagement. Currently, for example, higher income individuals and those who have been to college are far more engaged than others. Similarly, studies find that academically successful, white, and well-off individuals are far more likely than others to be provided with varied civic learning opportunities in school. Attending to these inequities is critical, he emphasized, noting that the premise of democracy is that all citizens have a voice, so if some groups are participating much more actively than others, this diminishes a democratic system.

MEDIA USE

In 2009, the typical U.S. youth between the ages of 8 and 18 devoted 7 hours and 38 minutes daily to using media (TV, audio, computers, video games, print, movies), noted Donald Roberts, and this was a full hour more time than young people had spent using media just 5 years previously. Because young people increasingly use several different media concurrently (for example, watching TV while surfing the web and texting a friend), they pack more than 10 hours of media use into those measured 7.5 hours.

There is no other waking activity that comes close to accounting for as much of young people's times, and adults are not far behind, he added. During the first 18 years of life, the typical U.S. youngster spends more time with just one of these media—television—than he or she spends in school. "American kids learn more from the media than they ever learn from school," Roberts observed. "It may not be what you want them to learn; it may not be what they need to be successful people, but every time they sit down and click or listen or watch, they're learning something."

Many studies have measured young people's television watching, Roberts explained, and a few have addressed movie viewing and radio listening, but there has been only one study, by the Kaiser Family Foundation, that used a representative sample to study young people's media use more broadly. Studying media use is complicated, he noted. Every time we go into the field," he pointed out, "somebody has invented a new technology." This expands the time available for media use: for example, new miniaturized and portable devices mean that time spent standing in line or sitting on a bus can also be an opportunity to pay a video game, communicate with friends, or watch a television show. It is generally possible to obtain measures of which and how many young people have which sorts of devices, but finer grained information is needed, and the data change very quickly. For example, knowing that the average household has more than two television sets, or that in 1999, 15 percent of 2- to 4-year-olds had a television in their bedroom, is not sufficient to understand the nature and extent of usage today.

At the same time, however, digital media also afford new ways to measure exposure, although they also raise very troublesome privacy issues as well. It is possible to record what sites young people are visiting, what they are doing, what shows they are watching, and other information, Roberts explained, but it is less clear how to understand

and code the sorts of content they are consuming and how to evaluate such findings. Looking just at media devoted to public affairs, he commented, it is difficult to identify precise boundaries for the category, given the range of programming and material that are available. "People really argue over this," he added, not only about perspectives on such programming as Fox news, but about what is learned through exposure to, say legal or police dramas that address civil rights and other civic issues.

Roberts noted that the Candian's Composite Learning Index (CLI) and the European Lifelong Learning Index (ELLI) both defined informal learning in relation to use of mass media. But he found their operational definitions of mass media to be limited—ELLI focused primarily on use of the internet and the CLI included use of the internet and reading materials. Roberts favors a more expansive definition that would include television and film, video games, and popular music, as well as other forms of electronic and print media. Entertainment media, he emphasized, are one of the primary sources of informal social learning for children and adolescents.

With these issues in mind, Roberts suggested two indicators and investigation of possibilities for a third. First, he suggested a basic measure of *access to media, beginning with measures of household spending on or ownership of media devices and sources.* Even better would be to supplement this measure with detailed looks at personal ownership (because usage might be very different for a computer or television in a child's room, for example, as opposed to one that is shared by several family members). Such data were collected by the Kaiser Family Foundation between 1999 and 2009, Roberts noted, but funding for that program was discontinued. He believes that an ongoing national survey of media habits would be valuable.

Roberts also advocated a measure of *media exposure that captures the nature of the content being consumed.* "Content matters," Roberts asserted, "and different young people focus on different sources and genres." Possible sources of data that could be used for this indicator include television and radio ratings services (e.g. Nielsen Ratings; Arbitron), sales or subscription data for age-targeted magazines and books, sales of age-targeted "educational" videogames, and tabulations of visits to internet destinations intended for young people. The most accurate data are likely to come from young users themselves, he added, which could also be obtained through a continuing national survey of media behavior.

Finally, Roberts speculated that an important outcome indicator for lifelong learning might be skill in critical thinking and problem solving. He thinks that there may be a relationship between the development of critical thinking and problem solving skills and media use, and he would like to see research in this area. He suggested that media use is important because it influences people's thinking and perceptions and because this influence is likely to vary across individuals, demographic groups, and time. Although no means of capturing such effects are currently available, he advocated that research on ways to explore such connections be a priority, since media use is increasingly pervasive. "If we are serious about capturing informal or incidental learning, then we need to think about all media, not just books and the internet, and we need to pay attention to what kinds of content are consumed." He argued that the evidence is clear that people use and respond to media and content differently as a function of age, so data are needed that capture exposure and content for people of all ages.

CULTURAL ACTIVITIES

Elizabeth Stage used the example of the Lawrence Hall of Science, a museum and science center run by the University of California at Berkeley, to highlight some key purposes for informal learning. Among the purposes that are often cited are personal fulfillment, civic engagement, workforce preparedness, and national competitiveness, but, Stage suggested, one could also view informal learning more broadly as a complement to formal learning. For her, the focus of study of informal learning should be on the opportunities it can provide, and the equity of those opportunities.

Informal learning differs from formal learning in a few ways, Stage noted. It is market based in the sense that it only occurs by choice. It tends to change faster than formal education can and to be organized around people rather than places. Increasingly, as digital media expand the possibilities, it can be customized to meet individual needs.

Lawrence Hall, she observed, is located on a high hill in the San Francisco Bay area and is not easily accessible to many city residents. Its mission, though, is to nurture science and mathematics learning for all, so, she noted "we have to go where the people are." As the staff have worked to make their programs accessible to everyone, Stage explained, they have found new ways to reach out to the public, including online activities and portable activities that can be taken to other locations.

Stage did not suggest specific indicators, but she identified two areas she believes are most important to explore. First, she argued that simple measures of participation, such as tracking how many people pass through a turnstile into a museum, are not particularly useful as indicators. More important, in her view, is to look at engagement, by, for example, observing and perhaps surveying small samples of people to understand in detail where and how they engaged in a particular learning experience and what they gained from it. Thus, an indicator might be one that broadly measures *engagement in learning opportunities outside of formal schooling, perhaps using surveys administered after learning experiences.* This approach is more difficult and expensive, she acknowledged, but in her view it offers an understanding of the new knowledge or attitudes gained, changes in behavior, or actions taken that could provide meaningful evidence of the value of informal education.

ISSUES IN INFORMAL LEARNING

Allan Collins raised a few points to launch the discussion of indicators of informal learning. First, he returned the discussion to the four pillars identified in the European Lifelong Learning Indicators: learning to know (the formal education system); learning to do (vocational training); learning to live together (learning for social cohesion); and learning to be (learning as personal growth).[28] Each of these pillars is assessed using a composite of subscales. For example, "learning to live together" covers participation in active citizenship; tolerance, trust, and openness; and inclusion in social networks. It measures such factors as work for voluntary or charitable organizations; membership in any political party; working in a political party or action group; holding the opinion that the country's cultural life is either enriched or undermined by

---

[28]See Hoskins, Cartwright, and Schoof (2010) for more information.

immigrants; holding the opinion that gays and lesbians should be free to live their own lives as they wish; having trust in other people; having interactions with friends, relatives, or colleagues; and having anyone to discuss personal matters with. Indicators of "learning to be" include participation in sports, attendance at or participation in cultural activities (ballet, dance, opera, concerts, and museums), participation in lifelong learning, personal use of the internet, internet access in the household, and accordance of working hours with family commitments. "If we want to make international comparisons," Collins noted, it might make sense to collect similar data.

Second, he stressed that fast-changing technological options are actually exacerbating the equity problems that are already an issue in education. He suggested that those who can afford them are eager consumers of new devices and services, but that others are increasingly left behind. It is critical, in his view, to find out how people are using new technology, but also to make a concerted effort to see that young children have equal access to the most important technological options.

Collins also suggested that schools and colleges currently "have a monopoly on certification," but that knowledge and skills gained in other settings can have as much value in the workplace. The more people learn outside of the K-16 setting, he suggested, the more important it is to think about assessing and certifying outside the system as well. This also is an equity issue, he pointed out, because alternative ways to demonstrate competence and skill may be of most benefit to students with fewer resources.

Much of the discussion focused on technology. Participants and presenters had several follow-up comments about its growing influence. The American Academy of Pediatrics still recommends that young children be limited to one hour or less of screen time per day, Roberts pointed out, and "there are social, economic, and intellectual implications from being glued to a screen all day." Some research has suggested that young people who have spent considerable time using screened devices "don't know how to read facial expressions anymore, and all the information we get from social interactions," he added. It will be important, in his view, to pay attention to possible negative outcomes as well as positive ones, in tracking the use of technology. It will also be important to keep an emphasis on measuring the content as well as the exposure, he reiterated, even though doing so requires value judgments.

A participant highlighted another distinction, between internet activity that is driven by friends and activity that is driven by interest. Real learning and involvement, this person suggested, is much more likely through activities that begin with substantive information as opposed to social contacts. Others questioned that distinction, noting that people seek out experiences not just on the internet but in other aspects of life, for all sorts of reasons, but can learn if they find substance in that experience.

Another participant noted the multinational and multilingual communities that are made possible on the internet, citing the example of Wikipedia. Such communities may be a new venue for lifelong learning, this person suggested, raising the question of whether there is a conceptual model guiding the discussion of indicators for this sort of learning. These sorts of interest-based groups, another responded, seem to be very strongly related to social engagement beyond the internet.

Some participants also pointed out that structured after-school and extracurricular programs had been overlooked in the presentations, but that they are have been shown to be an important influence, linked to performance in school and engagement with learning.

The National Center for Education Statistics has begun to examine extracurricular activities, one observer noted, and it will be important to learn more about when and where they are available, and about discrepancies in access to such opportunities, since they are often expensive. Another participant pointed out that studies of participation in extracurricular arts programs using data from the National Education Longitudinal Study have shown links with civic engagement.

The conversation closed with the observation that while it may be important to measure access to informal learning opportunities and participation in them, the point is not simply to have a sense of general quality of life. Learning outcomes, a participant suggested, should really be the focus.

7
Concluding Thoughts

The indicators suggested for each of the educational stages reflected many perspectives regarding which aspects of education are most important to track and how they can best be measured. The indicators suggested do not represent any consensus about what ought, ultimately, to be included in a system of education indicators, but do suggest the wide range of possibilities. The issues raised were the backdrop for a closing discussion in which five panelists, Emerson Elliott, Ronald Ferguson, Eugene García, Patricia Graham, and Marshall Smith, reflected on the task of selecting indicators for education. Each offered thoughts about the purposes for an indicator system and the criteria that should guide the selection, the structure or framework by which a set of education indicators might best be organized, their own candidates for the final list, and the issues they thought the presenters had overlooked.

PRIMARY PURPOSES FOR AN INDICATOR SYSTEM AND CRITERIA FOR SELECTION

The primary audience for an indicator system, in Emerson Elliott's view, is the general public. Researchers and policy makers may find them useful, he observed, but data that could be used in this sort of indicator system are not detailed enough to support decision making in schools, colleges, or classrooms. He suggested that effective indicators are based on three elements: research about measures, knowledge of what the intended audience needs, and understanding of how to communicate effectively with that audience.[29] In Elliott's view, the essential topics for national education indicators are (1) individual outcomes and behaviors and (2) providers and public investments.

For Ronald Ferguson the overarching purpose of an indicator system is to influence people's priorities. Some knowledge and skills tend to be easy to measure, he observed, and those received the most attention at the workshop. But more amorphous attitudes, values, goals, dispositions, and mindsets that are more difficult to measure are also important, he observed. He also noted that an indicator system should distinguish between in-school and out-of-school learning: he noted his particular interest in the ways that young people use their discretion in choosing out-of-school learning. A third priority for him is a system that tracks not only the availability of positive learning experiences, but also exposure to such stressors as poverty and such social toxins as violence and

---

[29]Elliott credited Connie Citro, director of the Committee on National Statistics of the National Academy of Sciences, for this observation.

60

substance abuse. He emphasized that if a major purpose of having an indicator system is to influence learning outcomes, then measures of the experiences that help to produce the outcomes are as important as the outcomes themselves.

Eugene García identified as the primary purpose of an indicator system informing the American public about the educational well-being of all and revealing the ways learning is associated with other indices of well-being (such as health, economic status, etc.). Such information could be used to evaluate current policy and guide new policies and, by identifying gaps and trends, highlight domains that may need specific policy and practice attention. It could also be used for international comparisons. García highlighted the importance of making sure that learning outcome measures are fleshed out with rich qualitative and contextual information. Such information may be difficult and costly to obtain, he acknowledged, but in his view it would significantly enhance understanding of the nation's educational well being. He also suggested that longitudinal measures are more informative than cross-sectional assessments alone. Longitudinal assessment allows for more robust analysis of individual and group progress, or lack thereof, over a specified period, he observed.[30]

"We are much better at static measures than at ones likely to stimulate positive change," suggested Patricia Graham. The easiest approach is to focus on educational institutions and their role in helping students learn, using measures of academic achievement. Assessments of academic achievement provide useful information, in her view, but she also asked "how raising test scores fits into the broader purposes of schooling," for which the metrics are less obvious. Expectations of U.S. schools have changed markedly in the past century, she noted, and promoting academic achievement for all is actually "a new assignment for them." A century ago, public schools were expected primarily to assimilate immigrant children into American society. In the ensuing decades other goals were added: fostering social adjustment and creativity; desegregating public institutions; and creating special programs for the poor, the disabled, the gifted, and English language learners, for example. It was only toward the end of the 20th century that the primary emphasis became high achievement for all students.

The available metrics have been very useful, Graham noted, in revealing obstacles to meeting that ambitious goal—achievement gaps and inequities in the education system. But achievement tests, in her view, should not be the sole, or even the principal indicators of learning. "We need more and better indicators of intellectual initiative, cooperative learning with others, and the ability to generate and assess new ideas and processes," she suggested. Also important, she added, are indicators of what students have learned regarding "their role in this democracy of majority rule and minority rights, and their capacity to respect others, play fair, and to learn the traditional adult virtues of hard work, accuracy, and responsibility." These attributes are much broader than simply getting and keeping a job, which is viewed as a key goal of schooling, but are nevertheless consistent with that desirable outcome, she noted. "We need metrics that reflect the education we

---

[30]In a discussion of longitudinal indicators, a participant emphasized that they are most suitable for tracking such benchmarks as graduation rates, where there is a stark difference between achieving it and not achieving it. When tracking the percentages of students who meet particular standards, for example, this person argued, the result should be treated as a continuous measure, not as a longitudinal benchmark.

really want our population to have, and highlight the conditions or procedures that are likely to [produce] that result," she concluded.

Marshall Smith noted that the United States has a complex and loosely structured education system and such systems are difficult to change through policy. The diffuse policy authority results in reduced coherence and predictability, he suggested. It is possible for new ideas to permeate such a system, but the knowledge base in education has often suffered from weak credibility. A respected indicator system, in his view, could provide useful leverage for innovation and improvement. He identified several specific goals for the indicator system:

- Each indicator should be made up of multiple measures or statistics, each of which is an important component. For example, academic growth for students from 6 to 18 could be one indicator and might include results from NAEP at 4th grade, PISA at age 15, and college admissions tests between ages 15 and 18.
- The set of indicators should reflect a vision of the future, should tell a story, and should be very easy to understand.
- The indicators should be flexible and able to change subtly over time as conditions change (in schools, the workplace, and with technology of all sorts, for example). Statistical procedures could be used to maintain trends)
- Whenever possible, the indicators should be aligned with international indicators so that comparisons can be made.

INDICATOR SYSTEM STRUCTURE

The way in which the indicator system is ultimately structured will reflect the goals it is designed to serve and, as Elliott observed, those goals have not yet been set. The indicators that were used to monitor progress toward the National Education Goals established in 1990, in Elliott's view, illustrate the importance of considering the goals carefully. Among those goals was that U.S. students would be first in the world in mathematics and science achievement by 2000, he noted, but "the effort lost steam as it became clear that action was not being taken to reach the goals."

Elliott noted that the workshop demonstrated that the presenters and participants each brought their own values to the question of what is most important to measure and also that rapidly changing contexts will need to be taken into account. There are at least four models to consider, Elliott added: the committee's framework: the Lifelong Learning indicators (ELLI) (see Chapter 6); the broad goals defined by the European Union (Economic Security, Social Cohesion, and Sustainability (Bertelsman Stiftung, 2008); and the four strands suggested by Marshall Smith, which were:

1. Human Outcomes—indicators of learning and doing;
2. Byproduct outcomes of learning, both for individuals and for society—research, innovation, the arts (music, dance, painting etc.), literature, sports, tolerance for others, etc.;
3. Formal infrastructure for learning—quality and availability of, for example, preschools, schools, certification systems, and formal networks; and
4. informal infrastructure for learning—quality and availability of supports within

the family and neighborhood, civil society, (social capital), structured learning on the job, independent and collective use of technology.

García observed that the steering committee's framework was structured around the current organization of formal education opportunities in the United States and that this may not be the best way to assess the overall well-being of lifelong learning processes. An alternative would be to use age as the organizing structure, so that indicators would be used to ask how well a particular age group is faring with respect to learning opportunities and outcomes in a variety of venues. In his view, this approach would fit more naturally with the way learning develops over time and would also align better with international indices that use age as the fundamental benchmark for comparisons.

ADDITIONAL CANDIDATES FOR A FINAL LIST

The five panelists identified a few ideas they believed had been overlooked by the presenters (see Table 7-1). Elliott listed specific opportunities as well as a few key issues. He noted that school and college data systems include longitudinal data that could be mined for a national indicator system, and also that there are examples of performance measures that could be useful models for U.S. indicators. He was concerned that three important issues—public school financing, equity issues, and the cost of higher education—were not well addressed, and also that the discussion did not pay sufficient heed to the importance of international comparisons. He also observed that the K-12 indicators did not address what happens to students after graduation from high school. Another gap was technology—though it was discussed at several points in the workshop it did not show up in the indicators for K-12 or postsecondary education.

Ferguson stressed the importance of making sure that data could be grouped by age, race, and socio economic status (as did many others throughout the discussion). He also noted the importance of examining trends and disparities, and particularly patterns of relationships within and across topic or issue categories. García stressed that the measures must be inclusive, reliable, valid, adaptable over time, and fiscally and practically feasible. It is also important, he added, that they have "face validity," meaning that they should be simple enough to be understood, even though they may be statistically complex and may capture complex phenomena.

Graham acknowledged the importance of existing measures of academic achievement, but added that measures of school conditions that support learning and achievement are also very important. It is also true that "school or college may not be the primary activity of youth," she suggested. Educational institutions are "in deep competition with considerably more compelling aspects of youth culture, especially technological enhancements and amusements," she observed. It will be important to understand them and also to understand how they can be utilized to enhance education, in her view. Graham also observed that economists have contributed much to understanding of learning and schooling, but that anthropology, sociology, and other disciplines will be needed to fully explore new questions about what explains past successes and what will make improvement more likely.

García also was concerned that most existing measures focus on specific outputs of educational interventions, and that greater attention is needed to the circumstances and contexts in which learning takes place. He emphasized the importance of attention to both language and immigration issues, noting that indicators of fluency in English and in other languages would be important not only for preschool years, but also throughout the K-12 experience. He noted as well that measures for the early childhood years as well as the later years of the life cycle are significantly more limited than the measures available for K-12 and higher education.

In closing, Chris Hoenig reemphasized the importance of the effort. He noted that the rapid pace of innovation and technological advances is continually reshaping education, both nationally and globally. As a result, the selection of key national indicators must take into account current conditions as well as anticipated future conditions. He thanked the steering committee and workshop participants for a thoughtful exploration of the issues and for providing the critical first step toward accomplishing this goal.

TABLE 7-1 Additional Indicators Suggested by the Final Panel

Emerson Elliott	<ul><li>Attendance, progression, and completion—for all levels of education</li><li>Readiness for each subsequent level</li><li>Classroom climate (preschool and K-12)</li><li>Spending at all levels, but per-pupil spending not sufficient. The metrics should capture state effort in relation to state GDP, distribution to individual schools distribution of benefits for college across income group.</li><li>Placement or other indicator of outcomes after high school or after college</li><li>Teacher evaluation and/or measure of professional status and progress (perhaps associated with school climate, as part of a composite indicator)</li><li>Learning: NAEP and international comparisons; additional modifications to capture critical thinking and problem-solving skills. Also include measures of knowing how to learn, if possible, and measures of practical application of learning (as in PISA)</li><li>College remediation</li><li>College access and individual student knowledge of access and ways to match the options with each students' own interests</li><li>College affordability and debt</li><li>Preschool NAEP-like assessment</li><li>College NAEP-like assessment</li><li>Civic engagement measures, such as voting and volunteering in relation to education</li></ul>
Ronald Ferguson	<ul><li>Learning outcomes (skill, knowledge, and orientations)</li><li>Measures of skill and knowledge (basic skills, critical thinking, other categories); also include measures of mindsets, dispositions/values/goals, "ideas about possible selves"</li></ul>

	• School-based learning experiences • Measures of quantity and quality of key inputs (teaching quality, peer supports, parenting quality) and methods (e.g. curricula, resources such as technology) • Out-of-school learning experience, learning, and lifestyles or behaviors, such as leisure reading, civic engagement, media use for news, media for informal learning, social networking, consumption of the news, habits of knowledge sharing, including access and home learning resources • Contexts and opportunities (availability of resources) • Exposure to stressors, accumulated social toxins, such as poverty
Eugene García	• Issues related to the diversity of the U.S. population—an index that captures demographic change and related assets and vulnerabilities • Measures of educational performance such as NAEP and longitudinal studies (e.g., ECLS-B, ECLS-K, and High School and Beyond) • Qualitative measures that assess the opportunity to learn—context measures would address issues of quality that could be tied to outcome measures
Patricia Graham	• Measures of academic achievement, e.g., NAEP and state standardized tests • Measures of school conditions, especially effective teaching • The role of technology in young people's lives
Marshall Smith	Indicators of human outcomes: learning and doing • Ages 0-6: quality of parent and family support for learning, including physical health and school readiness • Ages 6-18: academic growth (assessments, attainment/graduation); participation in community, students' belief that have "learned how to learn" and enjoy it • Ages 18-35: attendance and graduation tertiary education; participation in civil society (voting, networks, coaching) • Ages 25 and up: evidence of continued learning (formal, occupational, and informal) and of participation in civil society Indicators of nonacademic outcomes (byproducts of the system) • Ages 0-6: opportunity to participate in arts, sports, and group activities. • Ages 6-18: opportunity to participate in arts, sports, and group activities; experience with research, evidence, and research practice; and opportunity to be innovative in supportive environments. • Age18 and up: opportunity to create and use research and

	evidence in decision making, innovation, everyday work.
	Exposure to and participation in arts, literature, group work.
	Indicators for infrastructure for formal education system
	• Research productivity of institutions
	• For institutions and national, state, and local systems:
	   o quality, including degree of innovation, commitment to society, and fostering of learning
	   o costs for each participant in the system (e.g. students, taxpayers, government), and measures of efficiency
	   o availability and opportunity (including technology use to reach new students)
	Indicators for infrastructure of nonformal education system and employment training
	• Ages 0-6: quality of family/community support and learning systems
	• Ages 6-18: opportunities for extra support in school work (after school classes, summers, tutors). Use of study groups and networks, peers, web based tools and other independent study aides.
	• Ages 18 and up: opportunities for and utilization of employment training, engagement in nonformal learning networks, clubs, professional associations, community, web-based tools

References

Alamprese, J. (2012). *Indicators for adult education.* Paper presented at the NRC Workshop on Key National Education Indicators, National Research Council, January 27, Washington, DC.

Alexander, K.L., and Entwisle, D.R. (1988). Achievement in the first 2 years of school: Patterns and processes. *Monographs of the Society for Research in Child Development, 53*(2), 1-157.

Allen, J.P., Pianta, R.C., Gregory, A., Mikami, A.Y., and Lun, J. (2011). An interaction-based approach to enhancing secondary school instruction and student achievement. *Science, 333*(6045), 1034-1037.

Allensworth, E. (2012). *K-12 key education indicators.* Paper presented at the NRC Workshop on Key National Education Indicators, National Research Council, January 27, Washington, DC.

Arnold, D.S., Lonigan, C.J., Whitehurst, G.J., and Epstein, J.N. (1994). Accelerating language development through picture-book reading: Replication and extension to a videotape training format. *Journal of Educational Psychology, 86,* 235-243.

Auerbach, E.R. (1989). Toward a social-contextual approach to family literacy. *Harvard Educational Review, 59,* 165-181.

Barnett, W.S. (2011). Effectiveness of early educational intervention. *Science, 333,* 975-978.

Barnett, W.S. (2012). *Recommendations for preschool indicators.* Paper presented at the NRC Workshop on Key National Education Indicators, January 27, National Research Council, Washington, DC.

Barnett, W.S., Epstein, D.J., Carolan, M., Fitzgerald, J., Ackerman, D.J., and Friedman, A. (2010). *The state of preschool 2010: State preschool yearbook.* New Brunswick, NJ: NIEER.

Bertelsmann Stiftung. (2008). *European Lifelong Indicators (ELLI): Developing conceptual framework.* Working Paper Version 2.0, October. Available:

http://www.bertelsmann-stiftung.de/cps/rde/xbcr/SID-25D2C1C9-28106968/bst/ConceptualFramework_081022.pdf.

Bierman, K.L., Nix, R.L., Greenberg, M.T., Blair, C., and Domitrovich, C.E. (2008). Executive functions and school readiness intervention: Impact, moderation, and mediation in the Head Start REDI program. *Development and Psychopathology, 20* 21-843.

Bornstein, M.C., and Bradley, R.H. (Eds.). (2003). *Socioeconomic status, parenting, and child development.* Mahwah, NJ: Lawrence Erlbaum.

Bost, K.K., Vaughn, B.E., Washington, W.N., Gielinski, K.L., and Bradbard, M.R. (1998). Social competence, social support, and attachment: Demarcation of construct domains, measurement, and paths of influence for preschool children attending Head Start. *Child Development, 69,* 192-218.

Bradburn, N.M., Fuqua, C.J.E. (2010). Indicators and the federal statistical system: An essential but fraught partnership. *ANNALS of the American Academy of Political and Social Science, 630,* 89-108.

Bradley, R.H., Burchinal, M.R., and Casey, P.H. (2001). Early intervention: The moderating role of the home environment. *Applied Developmental Science, 5*, 2-8.

Bradley, R.H., Corwyn, R.F., Burchinal, M., McAdoo, H.P., and Coll, C.G. (2001). The home environments of children in the United States. Part II: Relations with behavioral development through age thirteen. *Child Development, 72*, 1868-1886.

Brice-Heath, S. (1986). Critical factors in literacy development. In S. de Castell, A. Luke, and K. Egan (Eds.), *Literacy, society, and schooling* (pp. 209-229). Cambridge, MA: Cambridge University Press.

Brooks-Gunn, J. and Markman, L.B. (2005). The contribution of parenting to ethnic and racial gaps in school readiness. *The Future of Children, 15*, 139-168.

Brown, J.L., Jones, S.M., LaRusso, M.D., and Aber, J.L. (2010). Improving classroom quality: Teacher influences and experimental impacts of the 4Rs program. *Journal of Educational Psychology, 102,* 153-167.

Bryk, A.S., and Hermanson, K.L. (1993). Educational indicator systems: Observations on their structure, interpretation, and use. *Review of Research in Education, 19*, 451-484.

Burchinal, M.R. (2012). *Key preschool variables to be included in a national system of education indicators.* Paper presented at the NRC Workshop on Key National Education Indicators, National Research Council, January 27, Washington, DC.

Burchinal, M.R., Campbell, F.A., Bryant, D.M., Wasik, B.H., and Ramey, C.T. (1997). Early intervention and mediating processes in cognitive performance of children of low-income African American families. *Child Development, 68*, 935-954.

Burchinal, M., Kainz, K., and Cai, Y. (2011). How well do our measures of quality predict child outcomes? A meta-analysis and coordinated analysis of data from large-scale studies of early childhood settings. In M. Zaslow (Ed.) *Reasons to take stock and strengthen our measures of quality.* Baltimore, MD: Brookes.

Burchinal, M., McCartney, K., Steinberg, L., Crosnoe, R., Friedman, S.L., McLoyd, V., and NICHD Early Child Care Research Network (2011). Examining the black-white achievement gap among low-income children using the NICHD Study of Early Child Care and Youth Development. *Child Development, 82*(5), 1404-1420.

Bus, A.G., and van IJzendoorn, M.H. (1988). Mother-child interactions, attachment and emergent literacy: A cross-sectional study. *Child Development, 59,* 1262-1272.

Buysse, V., and Peisner-Feinberg, E. (2010). Recognition and response: Response to intervention for pre-K. *Young Exceptional Children, 13,* 2-13.

Camilli, G., Vargas, S., Ryan, S., and Barnett, W.S. (2010). Meta-analysis of the effects of early education interventions on cognitive and social development. *Teachers College Record, 112*(3), 579-620.

Carnevale, A. P., Smith, N., and Strohl, J. (2010, June). *Help wanted: Projections of jobs and education requirements through 2018.* Washington, DC: Georgetown University Center on Education and the Workforce.

Caspe, M., and Lopez, E.M. (2006). *Lessons from family-strengthening interventions: Learning from evidence-based practice.* Cambridge, MA: Harvard Family Research Project.

Chazan-Cohen, R., Raikes, H., Brooks-Gunn, J., Ayoub, C., Pan, B.A., Kisker, E.E., Roggman, L.A., and Fuligni, A.S. (2009). Contributions of the parenting environment during the first five years of life to child outcomes at kindergarten entry in a low-income sample. *Early Education and Development, 20*, 958-977.

Chetty, R., Friedman, J.N., and Rockoff, J.E. (2011). *The long-term impacts of teachers: Teacher value-added and student outcomes in adulthood.* Working Paper 17699. Cambridge, MA: National Bureau of Economic Research.

Clements, D., and Sarama, J. (2008). Experimental evaluation of the effects of research-based preschool mathematics curriculum. *American Educational Research Journal, 45,* 443-494.

Committee on Measures of Student Success. (2011). *A report to Secretary of Education Arne Duncan.* Washington, DC: Author.

Delli Carpini, M., and Keeter, S. (1996). *What Americans know about politics and why it matters.* New Haven: Yale University Press.

Dickinson, D.K. (2011). Teachers' language practices and academic outcomes of preschool children. *Science, 333,* 964-967.

Dougherty, K.J. (2012). *Higher education indicators.* Paper presented at the NRC Workshop on Key National Education Indicators, January 27, National Research Council, Washington, DC.

Duncan, G., and Brooks-Gunn, J. (2000). Family poverty, welfare reform and child development. *Child Development, 71,* 188-196.

Duncan, G.J., Dowsett, C., Claessens, A., Magnuson, K., Huston, A., Klebanov, P., et al. (2007). School readiness and later achievement. *Developmental Psychology, 43,* 1428-1446.

Dynarski, M. (2012). *Suggested indicators for the K-12 institutional context.* Paper presented at the NRC Workshop on Key National Education Indicators, January 27, National Research Council, Washington, DC.

Early, D., Maxwell, K., Burchinal, M., Alva, S., Bender, R., Bryant, D., et al. (2007). Teachers' education, classroom quality, and young children's academic skills: Results from seven studies of preschool programs. *Child Development, 78*(2), 558-580.

Elliott, Emerson, J. (2009, December). *Precedents for education indicators.* Unpublished paper commissioned by State of the USA.

Espinosa, L.M. (2002). The connections between social-emotional development and literacy. *Kauffman Early Education Exchange, 1,* 31-44.

Essex, M.J., Klein, M.H., Miech, R., and Smider, N.A. (2001). Timing of initial exposure to maternal major depression and children's mental health symptoms in kindergarten. *British Journal of Psychiatry, 179,* 151-156.

Fantuzzo, J.W., Gadsden, V.L., and McDermott, P. A., (2010). An integrated curriculum to improve mathematics, language, and literacy for Head Start children. *American Educational Research Journal, 48,* 763-793.

Foster, M., Lambert, R., Abbott-Shim, M., McCarty, F., and Franze, S. (2005). A model of home learning environment and social risk factors in relation to children's emergent literacy and social outcomes. *Early Childhood Research Quarterly, 20,* 13-36.

García, T.I., and L'Orange, H.P. (2010). *Strong foundations: The state of state postsecondary data systems*. Boulder, CO: State Higher Education Executive Officers.

Gates Foundation, Bill & Melinda. (2010). *Learning about teaching: Initial findings of the Measures of Effective Teaching Project*. Available: http://www.metproject.org/downloads/Preliminary_Findings-Research_Paper.pdf.

Gates Foundation, Bill & Melinda. (2012). *Gathering feedback for teaching combining high-quality observations with student surveys and achievement gains.* Available: http://www.metproject.org/downloads/MET_Gathering_Feedback_Research_Paper.pdf.

Gill, B. (2012). *National indicators for K-12 education.* Paper presented at the NRC Workshop on Key National Education Indicators, January 27, National Research Council, Washington.

Ginsburg, A., Noell, J. and Plisko, V. (1988). Lessons learned from the wall chart. *Educational Evaluation and Policy Analysis, 10*(1), 1-12.

Gormley, Jr., W.T., Gayer, T., Phillips, D., Dawson, B. (2005). The effects of universal pre-K on cognitive development. *Developmental Psychology, 41*(6), 872-884.

Greene, J.P. (2002). *High school graduation rates in the United States* (Revised April 2002). Manhattan Institute for Policy Research, Civic Report, November, 2001. Available: http://www.manhattan-institute.org/html/cr_baeo.htm.

Grolnick, W.S., and Slowiaczek, M.L. (1994). Parents' involvement in children's schooling: A multidimensional conceptualization and motivational model. *Child Development, 65*, 237-252.

Grodsky, E., Warren, J.R., and Felts, E. (2008). Testing and social stratification in American education. *Annual Review of Sociology, 34*, 385-404.

Grossman, A.W., Churchill, J.D., McKinney, B.C., Kodish, I.M., Otte, S.L., and Greenough, W.T. (2003). Experience effects on brain development: Possible contribution to psychopathology. *Journal of Child Psychology and Psychiatry, 44*, 33-63.

Halle, T., Forry, N., Hair, E., Perper, K., Wandner, L., Wessel, J., and Vick, J. (2009). *Disparities in early learning and development: Lessons from the Early Childhood Longitudinal Study–Birth cohort (ECLS–B).* Washington, DC: Child Trends.

Hamre, B.K., and Pianta, R.C. (2005). Can instructional and emotional support in the first grade classroom make a difference for children at risk of school failure? *Child Development, 76*(5), 949-967.

Hanushek, E.A., and Rivkin, S.G. (2010). Generalizations about using value-added measures of teacher quality. *American Economic Review: Papers and Proceedings, 100,* 267-271.

Hart, D., Donnelly, T.M., Youniss, J., and Atkins, R. (2007). High school community service as a predictor of adult voting and volunteering. *American Educational Research Journal, 44,* 197-219.

Hart, B., and Risley, T.R. (1995). *Meaningful differences in the everyday experience of young American children.* Baltimore, MD: Brookes.

Haskins, R., and Rouse, C. (2005). *Closing achievement gaps.* The future of children policy brief. Washington, DC: Brookings Institution Press.

Heckman, J.J., and LaFontaine, P.A. (2008). The declining American high school graduation rate: Evidence, sources, and consequences. Cambridge, MA: National Bureau of Economic Research. NBER Report Research Summary Number 1. Available: http://www.nber.org/reporter/2008number1/heckman.html.

Hentschke, G. C., Lechuga, V. M., and Tierney, W.G. (Eds.). (2010). *For-profit colleges and universities: Their markets, regulation, performance, and place in higher education.* Sterling, VA: Stylus.

Hood, M., Conlon, E., and Andrews, G. (2008). Preschool home literacy practices and Children's literacy development: A longitudinal analysis. *Journal of Educational Psychology, 100,* 252-271.

Horn, M.B. (2011). *Beyond good and evil: Understanding the role of for-profits in education through the theories of disruptive innovation.* Washington, DC: American Enterprise Institute.

Hoskins, B., Cartwright, F., and Schoof, U. (2010). *Making lifelong learning tangible! The European lifelong learning indicators index.* Berlin, Germany: Bertelsmann Stiftung.

Institute of Medicine. (2009). *State of the USA health indicators: Letter report.* Committee on the State of the USA Health Indicators, Board on Population Health. Washington, DC: The National Academies Press.

Jung, L.A. (2010). Identifying families' supports and resources. In R.A. McWilliam (Ed.), *Working with families of young children with special needs* (pp. 9-26). New York: Guilford.

Kahne, J., Feezell, J., and Lee, N. (2012). Digital media literacy education and online civic and political participation. *International Journal of Communication, 6,* 1-24.

Kane, T., and Staiger, D. (2012). *Gathering feedback for teaching: Combining high-quality observations with student surveys and achievement gains.* Seattle, WA: Bill & Melinda Gates Foundation.

Karoly, L.A., Ghosh-Dastidar, B., Zellman, G.L., Perlman, M., and Fernyhough, L. (2008). *Prepared to learn: The nature and quality of early care and education for preschool-age children in California.* Santa Monica, CA: RAND.

Knoche, L.L., Edwards, C.P., Sheridan, S.M., Kupzyk, K.A., Marvin, C.A., Cline, K.D., and Clarke, B.L. (in press). Getting ready: Results of a randomized trial of a relationship-focused intervention on parent engagement in rural Early Head Start. *Infant Mental Health Journal.*

Korbin, J., and Coulton, C. (1997). Understanding the neighborhood context for children and families: Combining epidemiological and ethnographic approaches. In J. Brooks-Gunn, G.J. Duncan, and J.L. Aber (Eds.), *Neighborhood Poverty, Vol. 2* (65-79). New York: Sage.

Kutner, M., Greenberg, E., Jin, Y., Boyle, B., Hsu, Y., and Dunleavy, E. (2007). *Literacy in everyday life: Results from the 2003 National Assessment of Adult Literacy.* NCES No. 2007-490. Washington, DC: National Center for Education Statistics, Institute for Education Sciences, U.S. Department of Education.

Landry, S.H., Smith, K.E., Swank, P.R., Assel, M.A., and Vellet, S. (2001). Does early responsive parenting have a special importance for children's development or is consistency across early childhood necessary? *Developmental Psychology, 37,* 387-403.

Lee, V., and Burkham, D. (2002). *Inequality at the starting gate: Social background differences in achievement as children begin school.* Washington, DC: Economic Policy Institute.

Lemke, M., and Gonzales, P. (2006, June). *Findings from the condition of education 2006: U.S. student and adult performance on International Assessments of Educational Achievement.* NCES No. 2006073. Washington, DC: National Center for Education Statistics, Institute for Education Sciences, U.S. Department of Education.

Levine, S.C., Suriyakham, L.W., Rowe, M.L., Huttenlocher, J., and Gunderson, E.A. (2010). What counts in the development of young children's number knowledge? *Developmental Psychology, 46*(5), 1309-1319.

Literacy Collaborative. (2009). *Research on the effects of literacy collaborative on teaching and student learning.* Available: http://www.literacycollaborative.org/research/.

Long, B.T., and Kurlaender, M. (2009). Do community colleges provide a viable pathway to a baccalaureate degree? *Educational Evaluation and Policy Analysis, 31*(1), 30-53.

Magnuson, K.A., Meyers, M.K., Ruhm, C.J., and Waldfogel, J. (2004). Inequality in preschool education and school readiness. *American Education Research Journal, 41*(1), 115-157.

Masten, A.S., and Coatsworth, J.D. (1998). The development of competence in favorable and unfavorable environments: Lessons from research on successful children. *American Psychologist, 53*, 205-220.

McDevitt, M., and Kiousis, S. (2004). *Education for deliberative democracy: The long-term influence of Kids Voting USA.* CIRCLE Working Paper No. 22. Washington DC: Center for Information and Research on Civic Learning and Engagement.

Mutz, D.C. (2006). *Hearing the other side: Deliberative versus participatory democracy.* Cambridge: Cambridge University Press.

National Center for Higher Education and Public Policy. (2008). *Measuring Up, 2008: The national report card on higher education.* San Jose, CA: Author.

National Mathematics Advisory Panel. (2008). *Foundations for success: The final report of the National Mathematics Advisory Panel.* Washington, DC: U.S. Department of Education, Office of Planning, Evaluation and Policy Development.

National Research Council (1985). *Indicators of precollege education in science and mathematics: A preliminary review.* S.A. Raizen and L.V. Jones (Eds.). Committee on Indicators of Precollege Science and Mathematics Education, Commission on Behavioral and Social Sciences and Education. Washington, DC: National Academy Press.

National Research Council (1988). *Improving indicators of the quality of science and mathematics education in grades K-12.* R.J. Murnane and S.A. Raizen (Eds.). Committee on Indicators of Precollege Science and Mathematics Education, Commission on Behavioral and Social Sciences and Education. Washington, DC: National Academy Press.

National Research Council. (2001) *Understanding dropouts: Statistics, strategies, and high-stakes testing.* Committee on Educational Excellence and Testing Equity. A. Beatty, U. Neisser, W.T. Trent, and J.P. Heubert (Eds.). Board on Testing and Assessment, Center for Education. Division of Behavioral and Social Sciences and Education. Washington, DC: National Academy Press.

National Research Council. (2010). *Getting value out of value-added: Report of a workshop.* H. Braun, N. Chudowsky, and J.A. Koenig (Eds.). Committee on Value-Added Methodology for Instructional Improvement, Program Evaluation, and Accountability. Center for Education, Division of Behavioral and Social Sciences and Education. Washington, DC: The National Academies Press.

National Research Council. (2011a). *High school dropout, graduation, and completion rates: Better data, better measures, better decisions.* R.H. Hauser and J.A. Koenig (Eds.), Committee for Improved Measurement of High School Dropout and Completion Rates: Expert Guidance on Next Steps for Research and Policy Workshop. Center for Education, Division of Behavioral and Social Science and Education. Washington, DC: The National Academies Press.

National Research Council. (2011b). *A plan for evaluating the District of Columbia's public schools.* Committee on the Independent Evaluation of DC Public Schools, Division of Behavioral and Social Sciences and Education. Washington, DC: The National Academies Press.

National Task Force on Early Childhood Education for Hispanics. (2007). *Para nuestros niños: Expanding and improving early education for Hispanics.* Available: http://ecehispanic.org/work/expand_MainReport.pdf.

NICHD Early Child Care Research Network. (2002). Early child care and children's development prior to school entry: Results from the NICHD Study of Early Child Care. *American Educational Research Journal, 39,* 133-164.

NICHD Early Child Care Research Network. (2005). A day in third grade: A large-scale study of classroom quality and teacher and student behavior. *The Elementary School Journal, 105,* 305-323.

Organisation for Economic Co-operation and Development. (2009). *Tertiary-level educational attainment for age group 25-64: As a percentage of the population of that age group.* Education: Key Tables from OECD, No. 3. Paris, France: Author.

Organisation for Economic Co-operation and Development. (2010). *Tertiary education entry rates: First-time entrants as a percentage of the population in the corresponding age group.* Education: Key Tables from OECD, No. 2. Paris, France: Author.

Pan, B.A., Rowe, M.L., Singer, J.D., and Snow, C.E. (2005). Maternal correlates of growth in toddler vocabulary production in low-income families. *Child Development, 76,* 763-782.

Pascarella, E.T., and Terenzini, P.T. (1991). *How college affects students: Findings and insights from twenty years of research.* San Francisco: Jossey-Bass.

Pascarella, E.T., and Terenzini, P.T. (2005). *How college affects student (Vol. 2): A third decade of research.* San Francisco: Jossey-Bass.

Peisner-Feinberg, E.S., Burchinal, M.R., Clifford, R.M., Culkin, M.L., Howes, C., Kagan, S.L., and Yazejiam, N. (2001). The relation of preschool child care quality to children's cognitive and social developmental trajectories through second grade. *Child Development, 72*(5)*,* 1534-1553.

Perna, L.W. (2012). *Key indicators for the higher education stage.* Paper presented at the NRC Workshop on Key National Education Indicators, January 27, National Research Council, Washington, DC.

Phelan, P., Davidson, A.L., and Yu, H.C. (1998). *Adolescents' worlds: Negotiating family, peers, and school.* New York: Teachers College Press.

Pianta, R.C., Barnett, W.S., Burchinal, M., and Thornburg, K.R. (2009). The effects of preschool education: What we know, how public policy is or is not aligned with the evidence base, and what we need to know. *Psychological Science in the Public Interest, 10,* 49-88.

Pianta, R.C., Belsky, J., Houts, R., Morrison, F., and the NICHD Early Child Care Research Network. (2007). Opportunities to learn in America's elementary classrooms. *Science, 315,* 1795-1796.

Pianta, R.C., Belsky, J., Vandergrift, N., Houts, R. and Morrison, F. (2008). Classroom effects on children's achievement trajectories in elementary school. *American Educational Research Journal, 45,* 365-397.

Pianta R., Hamre, B.L. and Laparo, K. (2008). *Classroom assessment scoring system.* Baltimore, MD: Brookes.

Pianta, R., Mashburn, A., Downer, J., Hamre, B., and Justice, J. (2007). Effects of web-mediated professional development resources on teacher–child interactions in pre-kindergarten classrooms. *Early Childhood Research Quarterly, 23,* 431-451.

Planty, M., and Carlson, D. (2010). *Understanding education indicators: A practical primer for research and policy.* New York: Teachers College Press.

Popkin, S.L., and Dimock, M.A. (1999). Political knowledge and citizen competence. In S.L. Elkin, and K.E. Soltan (Eds.). *Citizen competence and democratic institutions.* University Park, PA: Pennsylvania State University Press.

Powell, D.R., Diamond, K.E., Burchinal, M.R., and Koehler, M.J. (2010). Effects of an early literacy professional development intervention on Head Start teachers and children. *Journal of Educational Psychology, 102,* 299-312

Pungello, E.P., Iruka, I.U., Dotterer, A.M., Mills-Koonce, R., and Reznick, J.S. (2009). The effects of income, race, sensitive parenting and harsh parenting on receptive and expressive language development in early childhood. *Developmental Psychology, 45,* 544-557.

Raikes, H., Pan, B.A., Luze, G., Tamis-MeLonda, C.S., Brooks-Gunn, J., Constantine, J., Tarullo, L.B., Raikes, H.A., and Rodriguez, E.T. (2006). Mother-child book reading in low-income families: Correlates and outcomes during the first three years. *Child Development, 77*, 924-953.

Rao, H., Betancourt, L.M., Giannetta, J.M., Brodsky, N.L., Korczykowski, M., Avants, B.B., et al. (2010). Early parental care is important for hippocampal maturation: Evidence from brain morphology in humans. *Neuroimage, 49*, 1144-1150.

Raviv, T., Kessenich, M., and Morrison, F.J. (2004). A mediational model of the association between socioeconomic status and three-year-old language abilities: The role of parenting factors. *Early Childhood Research Quarterly, 19*, 528-547.

Richburg-Hayes, L. (2012). *Key national indicators workshop.* Paper presented at the NRC Workshop on Key National Education Indicators, January 27, National Research Council, Washington, DC.

Roberts, D.F. (2012). *Notes on informal learning.* Paper presented at the NRC Workshop on Key National Education Indicators, January 27, National Research Council, Washington, DC.

Roberts, E., Bornstein, M.H., Slater, A.M., and Barrett, J. (1999). Early cognitive development and parental education. *Infant and Child Development, 8*, 49-62.

Rosenbaum, J.E., Deil-Amen, R., and Person, A.E. (2006). *After admission: From college access to college success.* New York: Russell Sage.

Russell Sage Foundation. (1912). *A comparative study of public school systems in the forty-eight states.* New York: Author.

Sameroff, A.J., Seifer, R., Barocas, R., Zax, M., and Greenspan, S. (1987). IQ scores of 4-year-old children: Social-environmental risk factors. *Pediatrics, 79*, 343-350.

Sampson, R.J. (1992). Family management and child development: Insights from social disorganization theory. *Advances in Criminological Theory, 3*, 63-93.

Schroeder, C. M., Scott, T.P., Tolson, H., Huang. T.Y. and L.H. Lee, L.H. (2007). A meta-analysis of national research: Effects of teaching strategies on student achievement in science in the United States, *Journal of Research in Science Teaching, 44*, 1436-1460.

Sheridan, S.M. (2012). *Indicators for preschool education: Focus on systems and contexts.* Paper presented at the NRC Workshop on Key National Education Indicators, January 27, National Research Council, Washington, DC.

Sheridan, S.M., Knoche, L.L., Edwards, C.P., Bovaird, J A., and Kupzyk, K.A. (2010). Parent engagement and school readiness: Effects of the Getting Ready intervention on preschool children's social-emotional competencies. *Early Education and Development, 21,* 125-156.

Sheridan, S.M., Knoche, L.L., Kupzyk, K.A., Edwards, C.P., and Marvin, C. (2011). A randomized trial examining the effects of parent engagement on early language and literacy: The getting ready intervention. *Journal of School Psychology, 49,* 361-383.

Smith, C., and Gomez, R. (2011). *Certifying adult education staff and faculty.* New York: Council for the Advancement of Adult Literacy.

Smith, M.S. (1988). Educational indicators. *The Phi Delta Kappan, 69*(7), 487-491.

Smith, M.S. (2012). *Thoughts on indicators.* Paper presented at the NRC Workshop on Key National Education Indicators, January 27, National Research Council, Washington, DC.

Snow, C.E. (1988). The last word: Questions about the emerging lexicon. In M.D. Smith and J. Locke (Eds.), *The emergent lexicon: The child's development of a linguistic vocabulary* (pp. 341-353). New York: Academic Press.

Snow, C.E. (1991). The theoretical basis for relationships between language and literacy in development. *Journal of Research in Childhood Education, 6,* 5-10.

Special Study Panel on Education Indicators. (1991, September). *Education counts: An indicator system to monitor the nation's educational health.* A report to the acting commissioner of education statistics.

Sroufe, L.A. (1983). Infant-caregiver attachment and patterns of adaptation in preschool: The roots of maladaptation and competence. In M. Permutter (Ed.), *Educating all students in the mainstream of regular education.* Baltimore, MD: Brookes.

Stage, E. (2012). *Indicators for lifelong, informal learning.* Paper presented at the NRC Workshop on Key National Education Indicators, January 27, National Research Council, Washington, DC.

Stern, D. (2012). *Indicators for employment-based education and training.* Paper presented at the NRC Workshop on Key National Education Indicators, January 27, National Research Council, Washington, DC.

Tamis-LeMonda, C.S., and Bornstein, M.H. (2002). Maternal responsiveness and early language acquisition. *Advances in Child Development & Behavior, 29,* 89-127.

Terenzini, P.T. (2012). *Indicators of learning outcomes in higher education.* Paper presented at the NRC Workshop on Key National Education Indicators, January 27, National Research Council, Washington, DC.

Tierney, W.G. (2012). *Postsecondary indicators and for-profit higher education.* Paper presented at the NRC Workshop on Key National Education Indicators, January 27, National Research Council, Washington, DC.

Tomasello, M., and Farrar, M. (1986). Joint attention and early language. *Child Development, 57,* 1454-1463.

Tout, K., Starr, R., Soli, M., Moodie, S., Kirby, G., and Boller, K. (2010). *Compendium of quality rating systems and evaluations.* Prepared for the Office of Planning, Research, and Evaluation. Available: http://www.acf.hhs.gov/programs/opre/cc/childcare_quality/compendium_qrs/qrs_compendium_final.pdf.

Turnbull, A.P., Blue-Banning, M., Turbiville, V., and Park, J. (1999). From parent education to partnership education: A call for a transformed focus. *Topics in Early Childhood Special Education, 19,* 164-171.

Turnbull, A., Turnbull, H.R., Erwin, E., Soodak, L., and Shogren, K. (2011). *Families, professionals, and exceptionality: Positive outcomes through partnerships and trust* (6th ed.). Upper Saddle River, NJ: Merrill Prentice Hall.

U.S. Census Bureau. (1976). *STATUS: A monthly chartbook of social and economic trends.* ST76-4. Washington, DC: Economic Surveys Division of the Bureau of the Census, Federal Statistical System.

U.S. Census Bureau. (2010). *2010 American community survey.* Washington, DC: Author. Available: http://factfinder2.census.gov/faces/tableservices/jsf/pages/productview.xhtml?pid-ACS_10_1YR_S1501&prodType=table.

U.S. Department of Education. (2010). *State-administered adult education program: Program year 2008-2009 enrollment.* Washington, DC: Division of Adult Education and Literacy, Office of Vocational and Adult Education, Author.

U.S Department of Education. (2011, September). *Adult education and family literacy act of 1998: Annual report to Congress 2008-2009.* Washington, DC: Division of Adult Education and Literacy, Office of Vocational and Adult Education, Author.

U.S. Department of Education (2012, May). *Implementation guidelines: Measures and methods for the National Reporting System for Adult Education.* Washington, DC: Division of Adult Education and Literacy, Office of Vocational and Adult Education, Author.

U.S. Department of Education and Division of Adult Education and Literacy. (2010). *State-administered adult education program: Program year 2008-2009 enrollment.* Washington, DC: Author.

U.S. Department of Health, Education, and Welfare (1969). *Toward a social report.* Washington, DC: U.S. Government Printing Office.

U.S. Government Accountability Office. (2004). *Informing our nation: Improving how to understand and assess the USA's position and progress.* GAO-05-1. Washington, DC: Author. Available: http://www.gao.gov/new.items/d051.pdf.

U.S. Government Accountability Office. (2011). *Key indicator systems: Experiences of other national and subnational systems offer insights for the United States.* GAO-11-396. Washington, DC: Author.

Vandell, D.L. (2012). *Recommended early childhood indicators.* Paper presented at the NRC Workshop on Key National Education Indicators, January 27, National Research Council, Washington, DC.

Vandell, D.L., Belsky, U., Burchinal, M., Steinberg, L., Vandergift, N., and NICHD NICHD Early Child Care Research Network. (2010). Do effects of early child care extend to age 15 years? Results from the NICHD Study of Early Child Care and Youth Development. *Child Development, 81,* 737-756.

Verhoeven, L., van Leeuwe, J., and Vermeer, A. (2011). Vocabulary growth and reading development across the elementary school years. *Scientific Studies of Reading, 15,* 8-25.

Vernon-Feagans, L., Kainz, K., Amendum, S., Ginsberg, M., Wood, T., and Bock, A. (2012). Targeted reading intervention: A coaching model to help classroom teachers with struggling readers. *Learning Disability Quarterly, 35,* 102-114.

Warren, J.R. (2012). *Questions for the panelists.* Paper presented at the NRC Workshop on Key National Education Indicators, January 27, National Research Council, Washington, DC.

Weigel, D.J., Martin, S.S., and Bennett, K.K. (2006a). Contributions of the home literacy environment to preschool-aged children's emerging literacy and language skills. *Early Child Development and Care, 176,* 357-378.

Weigel, D.J., Martin, S.S., and Bennett, K.K. (2006b). Mothers' literacy beliefs: Connections with the home literacy environment and preschool children's literacy development. *Journal of Early Childhood Literacy, 6,* 191-211.

Werquin, P. (2010). *Recognising non-formal and informal learning: Outcomes, policies, and practices.* Paris, France: Organisation for Economic Co-operation and Development.

Wood, D. (2011). *What are adjustment disorders?* Kenmore WA: Mental Health Matters. Available: http://www.mental-health-matters.com/disorders/45-adjustment/65-what-are-adjustment-disorders.

Yosso, T., and Solorzano, D. G. (2005). Conceptualizing a critical race theory in sociology. In M. Romero and E. Margolis (Eds.), *The Blackwell companion to social inequalities* (pp. 117-146). New York: Wiley.

Zukin, C., Keeter, S., Andolina, M., Jenkins, K. and Dellii Carpini, M.X. (2006). *A new engagement? Political participation, civic life, and the changing American citizen.* Oxford, England: Oxford University Press.

Appendix A
Workshop Agenda

Workshop on Key National Education Indicators
Keck Center, Room 101
500 Fifth St., NW
Washington DC
January 27-28, 2012

<u>Friday, January 27</u>

9:30-10:00   Opening Session to Lay Out the Context and Goals for the Workshop

- **Welcome, Introductions, Context for the Workshop**
  Robert Hauser, Executive Director, DBASSE
  Chris Hoenig, Senior Advisor to the Presidents, NAS
- **Overview of the Agenda, Discussion of the Framework**
  David Breneman, University of Virginia, Steering Committee Chair
  Diana Pullin, Boston College, Steering Committee

10:00-12:15   Indicators for the K-12 Stage

<u>Moderators:</u>

Henry Braun, Boston College, Steering Committee
Diana Pullin, Boston College, Steering Committee

10:00 Panel Discussion
- Elaine Allensworth, Consortium on Chicago School Research, Steering Committee
- Mark Dynarski, Pemberton Research, Steering Committee
- Brian Gill, Mathematica
- Robert Pianta, University of Virginia
- Rob Warren, University of Minnesota, Steering Committee

11:00-11:15   Break

11:15   Moderated Discussion

12:15-1:15   Working Lunch

1:15-3:15   Indicators for the Higher Education Stage

Moderators:
David Breneman, University of Virginia, Steering Committee
Lisa Lynch, Brandeis University, Steering Committee

1:15   Panel Discussion
- Kevin Dougherty, Columbia University
- University Laura Perna, University of Pennsylvania
- Lashawn Richburg-Hayes, MDRC
- Pat Terenzini, Pennsylvania State University
- William Tierney, University of Southern California

2:15   Moderated Discussion

3:15-3:30   Break

3:30-5:15   Indicators for Other Postsecondary Education/Training: Panel Discussion
Moderators:
Allan Collins,
Lisa Lynch Northwestern University, Steering Committee, Brandeis University, Steering Committee

3:30   Panel Discussion
- Judy Alamprese, Abt Associates
- Marshall S. Smith, Carnegie Foundation for the Advancement of Teaching
- David Stern, University of California, Berkeley

4:15   Moderated Discussion

5:15   Adjourn for the Day

Saturday, January 28

8:30–10:30   Indicators for the Preschool Stage

Moderators:
Elaine Allensworth, Consortium for Chicago School Research, Steering Committee
Ana Sol Gutierrez, Maryland State Legislature, Steering Committee

8:30   Panel Discussion
- Steve Barnett, Rutgers and National Institute for Early Education Research

83

- Margaret Burchinal, University of North Carolina at Chapel Hill
- Sue Sheridan, University of Nebraska
- Deborah Vandell, University of California at Irvine

9:30   Moderated Discussion

10:30-10:45   Break

10:45-12:15   Indicators for Lifelong, Informal Learning

Moderators:
Allan Collins, Northwestern University, Steering Committee
Diana Pullin, Boston College, Steering Committee

Panel Discussion
- Joseph Kahne, Mills College
- Donald Roberts, Stanford University
- Elizabeth Stage, Lawrence Hall

11:30   Moderated Discussion

12:15-1:00   Working Lunch

1:00-3:00   Synthesis of Ideas

Moderator:
David Breneman, University of Virginia, Steering Committee

1:00   Panel Discussion
- Emerson Elliott, National Council for Accreditation of Teacher Education
- Ronald Ferguson, Harvard University
- Eugene García, Arizona State University
- Patricia Graham, Harvard University
- Marshall S. Smith, Carnegie Foundation for the Advancement of Teaching

2:00   Moderated Discussion
Committee members respond to panelists
Discussion with audience members

3:00-3:15   Concluding Comments

3:15   Adjourn Workshop

Appendix B

Workshop Participants

Members of the Workshop Steering Committee
David Breneman, University of Virginia (Chair)
Elaine Allensworth, Consortium for Chicago School Research
Henry Braun, Boston College
Allan Collins, Northwestern University
Mark Dynarski, Pemberton Research
Ana Sol Gutierrez, Maryland House of Delegates
Lisa M. Lynch, Brandeis University
Diana Pullin, Boston College
J. Rob Warren, University of Minnesota

Workshop Panelists
Judith Alamprese, Abt Associates
Steve Barnett, Rutgers University
Margaret Burchinal, University of North Carolina at Chapel Hill
Kevin Dougherty, Columbia University
Emerson Elliott, National Council for Accreditation of Teacher Education
Ron Ferguson, Harvard University
Eugene García, Arizona State University
Brian Gill, Mathematica
Patricia Graham, Harvard University
Joseph Kahne, Mills College
Laura Perna, University of Pennsylvania
Robert Pianta, University of Virginia
Lashawn Richburg-Hayes, MDRC
Donald Roberts, Stanford University
Sue Sheridan, Nebraska University
Marshall S. Smith, Carnegie Foundation for the Advancement of Teaching
Elizabeth Stage, Lawrence Hall
David Stern, University of California at Berkeley
Patrick Terenzini, Pennsylvania State University
Bill Tierney, University of Southern California
Deborah Vandell, University of California at Irvine

Workshop Attendees
Robert Bell, National Science Foundation
Sharon Boivin, U.S. Department of Education
Norman Bradburn, University of Chicago
Chris Chapman, U.S. Department of Education
Traci Cook, U.S. Department of Education
Janice Earle, National Science Foundation
Larry Feinburg, National Assessment Governing Board
Edith Gummer, National Science Foundation
Carrie Heath Phillips, Council of Chief State School Officers
Monica Herk, National Board for Education Sciences
Sunil Iyengar, National Endowment for the Arts
Donna Plasket, University of Virginia
Michael Ross, Education Researcher
Terris Ross, U.S. Department of Education
Carl Wieman, White House Office of Science and Technology Policy

National Reearch Council Staff
Alexandra Beatty, Senior Program Officer, BOTA
Stuart Elliott, Director of BOTA
Robert Hauser, Executive Director, DBASSE
Chris Hoenig, President and CEO, The State of the USA
Kelly Iverson, Senior Program Assistant, BOTA
Judy Koenig, Study Director, BOTA

Preschool Stage

W. Steven Barnett
1. An index of children's prenatal exposure to hazards (e.g., tobacco, alcohol, drugs, maternal stress)
2. An index of learning and development collected at age 3 and kindergarten entry
3. An index of the home learning environment or experiences
4. The percentage of young children receiving early care and education outside the home by age
5. An index or measure of the quality of preschool programs

Margaret Burchinal
1. An indicator of the quality of interactions in child-care programs: the extent to which interactions between caregivers and children are warm, responsive, and linguistically rich.
2. An indicator of the quality and appropriateness of curricula in child-care programs: that curricula have the appropriate scope and sequence and incorporate methods for monitoring progress.
3. An indicator of the quality of teaching practices in child-care programs: that programs use intensive coaching linked to the curriculum or to promote evidence-based teaching practices.
4. An indicator of readiness: that children acquire the language, academic, attention, and social skills to be ready for school.

Susan Sheridan
1. An index of the number/percent of families/households that provide attached parent-child relationships (e.g., display of affection, physical proximity, contingent positive reinforcement).
2. An index of the percent of families that provide enriching and stimulating home environments.
3. Percentage of families with multiple risk factors (such as, poverty, single adult household, household is non-English speaking in a predominantly English-speaking environment/community, parent(s) have less than a high school education, teen parent(s), parental mental illness).
4. A measure of access to support systems, such as the number of children/families receiving health, mental health, and social services.

5. Percentage of parents who are connected to or partner with early childhood providers via interactions that are planned and collaborative.

Deborah Vandell
1. An index of the type of child care available for children ages 0-3 and 3-6
2. An assessment of the quality of care provided to children in child care program
3. An assessment of learning readiness administered in kindergarten
4. Percentage of children (birth to age 5) living in poverty, especially living in persistent poverty during this period

K-12

Elaine Allensworth
1. Attendance in school by age
2. College readiness levels by age/grade
3. A measure of safe and orderly school climate
4. A measure of school culture (supportiveness) for college and careers
5. A measure of collaborative school community that focuses on student learning

Mark Dynarski
1. Parent satisfaction with education
2. K-12 education spending as a share of GDP
3. K-12 education spending per student
4. Teacher student ratio

Brian Gill
1. Proportion of teachers whose evaluations distinguish them from a basic standard, using measures of their contributions to student achievement and their professional practice.
2. Percentage of K-12 education funding spent on research and development
3. Voter registration rate of 18-21 year olds

Robert Pianta
1. Teachers with mastery-level and current knowledge of content they are teaching
2. Teachers with mastery-level and contemporary knowledge of child and adolescent development
3. Curricula with demonstrated evidence of associations with student learning
4. Teacher-student interactions that demonstrate high levels and qualities of involvement, stimulation, and expansion of thinking and cognition, and sensitivity to students' perspectives, individual experiences, and backgrounds
5. Teacher-student interactions that foster relationships with and among students
6. Opportunities to learn that challenge thinking

Rob Warren
1. Command of core content (NAEP scores)
2. Grade retention rates through 8th grade
3. High school completion rates
4. A measure of opportunity to learn
5. A measure of college readiness

Higher Education

Kevin Dougherty
1. Degree completion, particularly for community colleges
2. Measures of education progression (toward degree completion)
3. Job placement and earnings
4. Student learning
5. Societal outcomes, such as reduced income and wealth inequality
6. Student knowledge of the college access and success process

Laura Perna
1. Institutional completion/graduation rates
2. Educational attainment (10 years after first enrolled)
3. Affordability
4. Magnitude and manageability of debt
5. Economic benefits to individuals (employment rates and salaries)
6. College readiness and preparation
7. A measure of equity (distributions of students by key demographic characteristics across different types of postsecondary education institutions and higher education outcomes

Lashawn Richburg-Hayes
1. The proportion of first-time, first-year students that are college-ready at matriculation
2. The proportion of students who passed the developmental course requirements within three semesters, if not college-ready at matriculation
3. The cumulative total of degree-applicable credits earned, possibly relative to a benchmark of 60 credits (the average credits required for an associate's degree)

Patrick Terenzini
1. Higher-order cognitive skills (critical thinking, problem solving, synthesizing and evaluating evidence)
2. Occupational competence in some field
3. Civic awareness and responsibility
4. Global and inter-cultural competence
5. Moral reasoning

Bill Tierney
1. Ability to overcome remedial needs
2. Ability to retain and graduate students
3. Ability to provide skills necessary for successful careers
4. Ability to place students in jobs comparable to their education/training
5. Debt incurred is manageable in relationship to income

Other Postsecondary Education and Training

Judy Alamprese
1. How literate are adults in America?
2. Do adults who re-enter education after leaving high school without a diploma earn postsecondary credentials?
3. What are the civic and social outcomes of adult education?
4. What proportion of national wealth is spent on adult education?
5. How prepared are adult education instructors to facilitate adults' achievement of their education and learning goals?

David Stern
1. What types of on-the-job training are provided, including classroom instruction, work practices, and on-line resources
2. Extent to which employed individuals participate in on-the-job training
3. Measures of skill shortages in new or changing occupations/industries

Marshall Smith
1. Do U.S. employers promote and create learning environments within their organization? Do they support innovation? Do employees have collective learning plans?
2. What percentage of the adult population believes that they know how to learn and that they are motivated to exercise this skill?
3. Is there a healthy environment of nonschool institutions and services that support deliberate learning in areas associated with occupational interests?
4. Are people taking advantage of the sets of opportunities made available to them?

Lifelong, Informal Learning

Joseph Kahne
1. Political engagement
2. Civic engagement
3. Learning through engagement with civic and political information through media
4. Learning through engagement with diverse views on civic and political issues
5. Civic learning opportunities

Don Roberts
1. Measure of access, such as a measure of household spending/ownership of various media
2. Measure of exposure to media and types of media children are exposed to
3. Some measure of impact/influence, such as a measure of the relationship between use of various media platforms and critical-thinking/problem-solving skills

Elizabeth Stage
1. Participation in cultural activities, such as going to museums, exhibits, etc.
2. Engagement in cultural activities